T0235902

Lecture Notes in Computer Science 12402

More information about this series at http://www.springer.com/series/7409

Surya Nepal · Wenqi Cao ·
Aziz Nasridinov · MD Zakirul Alam Bhuiyan ·
Xuan Guo · Liang-Jie Zhang (Eds.)

Big Data – BigData 2020

9th International Conference
Held as Part of the Services Conference Federation, SCF 2020
Honolulu, HI, USA, September 18–20, 2020
Proceedings

 Springer

Editors
Surya Nepal
CSIRO
Marsfield, NSW, Australia

Aziz Nasridinov
Chungbuk National University
Cheongju, Korea (Republic of)

Xuan Guo ⓘ
University of North Texas
Denton, TX, USA

Wenqi Cao
Facebook (United States)
Menlo Park, CA, USA

MD Zakirul Alam Bhuiyan ⓘ
Fordham University
Bronx, NY, USA

Liang-Jie Zhang ⓘ
Kingdee International Software
Group Co. Ltd.
Shenzhen, China

ISSN 0302-9743   ISSN 1611-3349 (electronic)
Lecture Notes in Computer Science
ISBN 978-3-030-59611-8   ISBN 978-3-030-59612-5 (eBook)
https://doi.org/10.1007/978-3-030-59612-5

LNCS Sublibrary: SL3 – Information Systems and Applications, incl. Internet/Web, and HCI

This Springer imprint is published by the registered company Springer Nature Switzerland AG
The registered company address is: Gewerbestrasse 11, 6330 Cham, Switzerland

Preface

The International Congress on Big Data (BigData 2020) aims to provide an international forum that formally explores various business insights of all kinds of value-added services. Big Data is a key enabler for exploring business insights and economics of services.

BigData 2020 is a member of the Services Conference Federation (SCF). SCF 2020 had the following 10 collocated service-oriented sister conferences: the International Conference on Web Services (ICWS 2020), the International Conference on Cloud Computing (CLOUD 2020), the International Conference on Services Computing (SCC 2020), the International Conference on Big Data (BigData 2020), the International Conference on AI & Mobile Services (AIMS 2020), the World Congress on Services (SERVICES 2020), the International Conference on Internet of Things (ICIOT 2020), the International Conference on Cognitive Computing (ICCC 2020), the International Conference on Edge Computing (EDGE 2020), and the International Conference on Blockchain (ICBC 2020). As the founding member of SCF, the First International Conference on Web Services (ICWS 2003) was held in June 2003 in Las Vegas, USA. Meanwhile, the First International Conference on Web Services - Europe 2003 (ICWS-Europe 2003) was held in Germany in October 2003. ICWS-Europe 2003 was an extended event of ICWS 2003, and held in Europe. In 2004, ICWS-Europe was changed to the European Conference on Web Services (ECOWS), which was held in Erfurt, Germany. To celebrate its 18th birthday, SCF 2020 was successfully held in Hawaii, USA.

This volume presents the accepted papers for the BigData 2020, held as a fully virtual conference during September 18–20, 2020. The major topics of BigData 2020 included but were not limited to: Big Data Architecture, Big Data Modeling, Big Data as a Service, Big Data for Vertical Industries (Government, Healthcare, etc.), Big Data Analytics, Big Data Toolkits, Big Data Open Platforms, Economic Analysis, Big Data for Enterprise Transformation, Big Data in Business Performance Management, Big Data for Business Model Innovations and Analytics, Big Data in Enterprise Management Models and Practices, Big Data in Government Management Models and Practices, and Big Data in Smart Planet Solutions.

We accepted 18 papers, including 16 full papers and 2 short papers. Each was reviewed and selected by at least three independent members of the BigData 2020 International Program Committee. We are pleased to thank the authors whose submissions and participation made this conference possible. We also want to express our thanks to the Organizing Committee and Program Committee members for their dedication in helping to organize the conference and reviewing the submissions. We thank all volunteers, authors, and conference participants for their great contributions to the fast-growing worldwide services innovations community.

July 2020

Surya Nepal
Wenqi Cao
Aziz Nasridinov
MD Zakirul Alam Bhuiyan
Xuan Guo
Liang-Jie Zhang

Organization

General Chairs

Seung-Jong Jay Park	Louisiana State University, USA
Minyi Guo	Shanghai Jiaotong University, China
Min Li	Central South University, China

Program Chairs

Surya Nepal	CSIRO's Data61, Australia
Wenqi Cao	Facebook, Inc., USA
Aziz Nasridinov	Chungbuk National University, South Korea
MD Zakirul Alam Bhuiyan	Fordham University, USA
Xuan Guo	University of North Texas, USA

Services Conference Federation (SCF 2020)

General Chairs

Yi Pan	Georgia State University, USA
Samee U. Khan	North Dakota State University, USA
Wu Chou	Vice President of Artificial Intelligence & Software at Essenlix Corporation, USA
Ali Arsanjani	Amazon Web Services (AWS), USA

Program Chair

Liang-Jie Zhang	Kingdee International Software Group Co., Ltd, China

Industry Track Chair

Siva Kantamneni	Principal/Partner at Deloitte Consulting, USA

CFO

Min Luo	Huawei, USA

Industry Exhibit and International Affairs Chair

Zhixiong Chen	Mercy College, USA

Operations Committee

Jing Zeng	Yundee Intelligence Co., Ltd, China
Yishuang Ning	Tsinghua University, China
Sheng He	Tsinghua University, China
Yang Liu	Tsinghua University, China

Steering Committee

Calton Pu (Co-chair)	Georgia Tech, USA
Liang-Jie Zhang (Co-chair)	Kingdee International Software Group Co., Ltd, China

AIMS 2020 Program Committee

Shreyansh Bhatt	Amazon Web Services (AWS), USA
Kosaku Kimura	Fujitsu Laboratories Ltd, Japan
Harald Kornmayer	DHBW Mannheim, Germany
Ugur Kursuncu	University of South Carolina, USA
Sarasi Lalithsena	IBM Sillicon Valley, USA
Yu Liang	The University of Tennessee at Chattanooga, USA
Hemant Purohit	George Mason University, USA
Luiz Angelo Steffenel	Université de Reims Champagne-Ardenne, France
Pierre Sutra	Télécom SudParis, France
Raju Vatsavai	North Carolina State University, USA
Jianwu Wang	University of Maryland, Baltimore County, USA
Wenbo Wang	GoDaddy, USA

Conference Sponsor – Services Society

Services Society (S2) is a nonprofit professional organization that has been created to promote worldwide research and technical collaboration in services innovation among academia and industrial professionals. Its members are volunteers from industry and academia with common interests. S2 is registered in the USA as a "501(c) organization," which means that it is an American tax-exempt nonprofit organization. S2 collaborates with other professional organizations to sponsor or co-sponsor conferences and to promote an effective services curriculum in colleges and universities. The S2 initiates and promotes a "Services University" program worldwide to bridge the gap between industrial needs and university instruction.

The services sector accounted for 79.5% of the USA's GDP in 2016. The world's most service-oriented economy, with service sectors accounting for more than 90% of GDP. S2 has formed 10 Special Interest Groups (SIGs) to support technology and domain specific professional activities:

- Special Interest Group on Web Services (SIG-WS)
- Special Interest Group on Services Computing (SIG-SC)
- Special Interest Group on Services Industry (SIG-SI)
- Special Interest Group on Big Data (SIG-BD)
- Special Interest Group on Cloud Computing (SIG-CLOUD)
- Special Interest Group on Artificial Intelligence (SIG-AI)
- Special Interest Group on Edge Computing (SIG-EC)
- Special Interest Group on Cognitive Computing (SIG-CC)
- Special Interest Group on Blockchain (SIG-BC)
- Special Interest Group on Internet of Things (SIG-IOT)

About the Services Conference Federation (SCF)

As the founding member of the Services Conference Federation (SCF), the First International Conference on Web Services (ICWS 2003) was held in June 2003 in Las Vegas, USA. Meanwhile, the First International Conference on Web Services - Europe 2003 (ICWS-Europe 2003) was held in Germany in October 2003. ICWS-Europe 2003 was an extended event of ICWS 2003, and held in Europe. In 2004, ICWS-Europe was changed to the European Conference on Web Services (ECOWS), which was held in Erfurt, Germany. SCF 2019 was held successfully in San Diego, USA. To celebrate its 18th birthday, SCF 2020 was held virtually during September 18–20, 2020.

In the past 17 years, the ICWS community has been expanded from Web engineering innovations to scientific research for the whole services industry. The service delivery platforms have been expanded to mobile platforms, Internet of Things (IoT), cloud computing, and edge computing. The services ecosystem is gradually enabled, value added, and intelligence embedded through enabling technologies such as big data, artificial intelligence (AI), and cognitive computing. In the coming years, all the transactions with multiple parties involved will be transformed to blockchain.

Based on the technology trends and best practices in the field, SCF will continue serving as the conference umbrella's code name for all service-related conferences. SCF 2020 defines the future of New ABCDE (AI, Blockchain, Cloud, big Data, Everything is connected), which enable IoT and enter the 5G for the Services Era. SCF 2020's 10 collocated theme topic conferences all center around "services," while each focusing on exploring different themes (web-based services, cloud-based services, big data-based services, services innovation lifecycle, AI-driven ubiquitous services, blockchain driven trust service-ecosystems, industry-specific services and applications, and emerging service-oriented technologies). SCF includes 10 service-oriented conferences: ICWS, CLOUD, SCC, BigData Congress, AIMS, SERVICES, ICIOT, EDGE, ICCC, and ICBC. The SCF 2020 members are listed as follows:

[1] The International Conference on Web Services (ICWS 2020, http://icws.org/) is the flagship theme-topic conference for Web-based services, featuring Web services modeling, development, publishing, discovery, composition, testing, adaptation, delivery, as well as the latest API standards.

[2] The International Conference on Cloud Computing (CLOUD 2020, http://thecloudcomputing.org/) is the flagship theme-topic conference for modeling, developing, publishing, monitoring, managing, delivering XaaS (Everything as a Service) in the context of various types of cloud environments.

[3] The International Conference on Big Data (BigData 2020, http://bigdatacongress.org/) is the emerging theme-topic conference for the scientific and engineering innovations of big data.

[4] The International Conference on Services Computing (SCC 2020, http://thescc.org/) is the flagship theme-topic conference for services innovation lifecycle that includes enterprise modeling, business consulting, solution creation, services

orchestration, services optimization, services management, services marketing, and business process integration and management.

[5] The International Conference on AI & Mobile Services (AIMS 2020, http:// ai1000.org/) is the emerging theme-topic conference for the science and technology of AI, and the development, publication, discovery, orchestration, invocation, testing, delivery, and certification of AI-enabled services and mobile applications.

[6] The World Congress on Services (SERVICES 2020, http://servicescongress.org/) focuses on emerging service-oriented technologies and the industry-specific services and solutions.

[7] The International Conference on Cognitive Computing (ICCC 2020, http:// thecognitivecomputing.org/) focuses on the Sensing Intelligence (SI) as a Service (SIaaS) which makes systems listen, speak, see, smell, taste, understand, interact, and walk in the context of scientific research and engineering solutions.

[8] The International Conference on Internet of Things (ICIOT 2020, http://iciot.org/) focuses on the creation of IoT technologies and development of IoT services.

[9] The International Conference on Edge Computing (EDGE 2020, http:// theedgecomputing.org/) focuses on the state of the art and practice of edge computing including but not limited to localized resource sharing, connections with the cloud, and 5G devices and applications.

[10] The International Conference on Blockchain (ICBC 2020, http://blockchain1000. org/) concentrates on blockchain-based services and enabling technologies.

Some highlights of SCF 2020 are shown below:

- **Bigger Platform:** The 10 collocated conferences (SCF 2020) are sponsored by the Services Society (S2) which is the world-leading nonprofit organization (501 c(3)) dedicated to serving more than 30,000 worldwide services computing researchers and practitioners. Bigger platform means bigger opportunities to all volunteers, authors, and participants. Meanwhile, Springer sponsors the Best Paper Awards and other professional activities. All 10 conference proceedings of SCF 2020 have been published by Springer and indexed in ISI Conference Proceedings Citation Index (included in Web of Science), Engineering Index EI (Compendex and Inspec databases), DBLP, Google Scholar, IO-Port, MathSciNet, Scopus, and ZBlMath.
- **Brighter Future:** While celebrating the 2020 version of ICWS, SCF 2020 highlights the Third International Conference on Blockchain (ICBC 2020) to build the fundamental infrastructure for enabling secure and trusted service ecosystems. It will also lead our community members to create their own brighter future.
- **Better Model:** SCF 2020 continues to leverage the invented Conference Blockchain Model (CBM) to innovate the organizing practices for all the 10 theme conferences.

Contents

Application Track

Short Paper Track

Research Track

Entropy-Based Approach to Efficient Cleaning of Big Data in Hierarchical Databases

Eugene Levner[1(✉)], Boris Kriheli[1,2], Arriel Benis[1],
Alexander Ptuskin[3], Amir Elalouf[4], Sharon Hovav[4,5],
and Shai Ashkenazi[6]

[1] Holon Institute of Technology, 5810201 Holon, Israel
{levner,borisk,arrielb}@hit.ac.il
[2] Ashkelon Academic College, 78211 Ashkelon, Israel
[3] Kaluga Branch of Moscow Bauman Technical University,
248004 Kaluga, Russia
aptuskin@mail.ru
[4] Bar Ilan University, 52900 Ramat-Gan, Israel
amir.elalouf@biu.ac.ill
[5] Clalit Health Services, 62098 Tel Aviv, Israel
sharonh@clalit.org.i
[6] Ariel University, 40700 Ariel, Israel
shai.ashkenazi7@gmail.com

Abstract. When databases are at risk of containing erroneous, redundant, or obsolete data, a cleaning procedure is used to detect, correct or remove such undesirable records. We propose a methodology for improving data cleaning efficiency in a large hierarchical database. The methodology relies on Shannon's information entropy for measuring the amount of information stored in databases. This approach, which builds on previously-gathered statistical data regarding the prevalence of errors in the database, enables the decision maker to determine which components of the database are likely to have undergone more information loss, and thus to prioritize those components for cleaning. In particular, in cases where the cleaning process is iterative (from the root node down), the entropic approach produces a scientifically motivated stopping rule that determines the optimal (i.e. minimally required) number of tiers in the hierarchical database that need to be examined. This stopping rule defines a more streamlined representation of the database, in which less informative tiers are eliminated.

Keywords: Data cleaning · Entropy-based analytics · Entropy evaluation

1 Introduction

As datasets have grown vastly larger in recent years, the risk that a database might contain "invalid" data—i.e., data that are erroneous, redundant or obsolete—has also escalated. This trend highlights the need for effective procedures for *data cleaning*, broadly defined as the detection and removal of errors and redundant or obsolete information from datasets of various structures in order to improve their quality.

© Springer Nature Switzerland AG 2020
S. Nepal et al. (Eds.): BIGDATA 2020, LNCS 12402, pp. 3–12, 2020.
https://doi.org/10.1007/978-3-030-59612-5_1

In general, data cleaning entails three main processes, which have been discussed extensively in prior research (see, e.g. [1–4]): (i) identification of invalid records; (ii) validation, in which errors are corrected and invalid information is eliminated; and (iii) imputation, i.e., the replacement of invalid records.

In a very large database, it can be highly time-consuming and even computationally infeasible to execute each of these processes on the complete data structure. Herein, we develop a methodology for improving the efficiency of data cleaning in a hierarchical database (HDB)—the oldest and simplest type of database, in which the data are structured as a (tiered) hierarchical tree-type graph. In an HDB, data cleaning ultimately entails transforming the original database into a reduced graph model with a smaller set of tiers and nodes, in which only relevant information is retained.

The basic idea of our approach is to quantify the extent to which invalid data are likely to have caused information quantity loss in specific components of the database, thereby enabling the decision maker to prioritize these components for cleaning and to ignore components in which information loss is less severe. Though diverse definitions exist for information quantity in an HDB (see [5, 6] for in-depth discussions), we suggest that *entropy*—and specifically, Shannon's information entropy [7, 8]—can serve a useful tool for measuring the quantity of information in a database [9, 10] and for identifying where the most informative records are located. In general, entropy is a measure of uncertainty, chaos, and absence of knowledge. Thus, entropy taken with the opposite sign is equivalent to the information amount, or the knowledge available [7, 8].

To our knowledge, entropy-based approaches to information measurement and data cleaning in databases have received comparatively little attention in the literature. Yet, entropy is used in many domains as a means of evaluating system complexity (e.g., [11–16]). Several studies have explored the fact that a high level of entropy (or chaos) in a complex system impedes perfect system performance: the greater the entropy, the more uncertain is the state of the entire complex system and, consequently, a larger amount of information is required to monitor and manage the system. Adopting this perspective, herein we consider entropy as a complexity measure for estimating the amount of information contained in specific components of an HDB.

Prior studies [12, 15, 16] used entropy-based approaches in considering simple single-tier production systems. The present paper extends these approaches to the case of multi-layer hierarchical data structures, and, in particular, the approach suggested by Karp and Ronen [16]. However, in contrast to the prior works, we investigate entropy in databases rather than in production systems, and hence the contents and measurement of the entropy are different.

Though a key objective in developing our approach is to provide an efficient means of cleaning the data in an HDB (towards improving data quality), our methodology enables us to achieve a broader objective: producing a more streamlined representation of an HDB. Specifically, as elaborated in what follows, our approach can be used to identify which data tiers in an HDB contain essential amounts of information. In effect, the decision maker might choose to eliminate subsequent nonessential tiers, thereby reducing the size of the HDB without losing important information.

The remainder of the paper is structured as follows. In the next section we provide a basic overview of the HDB architecture. Section 3 provides basic definitions used in our entropy-based approach, and Sect. 4 introduces the approach itself. In Sect. 5, we briefly describe the application of our approach to two real-life databases. Section 6 concludes.

2 Overview of the HDB Architecture

An HDB architecture is a tree-like graph structure in which the data are stored as records, where a record is defined as a collection of fields [2–4]. Each node in the graph represents a cluster of records. The nodes are organized in tiers that are sequentially linked to one another through graph arcs. Specifically, the "top" tier—tier zero, the root node—contains all initial records in the database. The root is linked to a first tier of nodes, representing clusters of these records. Each of these nodes is linked to a second tier of sub-clusters, and so on. We study the HDBs by traversing the graph from the top down, starting from the root node.

HDB models are often used in situations in which the primary focus of information gathering is on a concrete system hierarchy, such as a list of components of a complex equipment or a company organization chart. Examples of such databases are shown in in Figs. 1 and 2, which, respectively, illustrate the hierarchical structures of a pharmaceutical materials database and an automobile component database. Specifically, Fig. 1 is an example of an HDB representing the distribution of pharmaceutical materials managed by a healthcare provider. The root node contains all records of all pharmaceutical materials that the provider (Distribution Center) holds. In the first tier, these records are clustered according to their distribution across the large geographical regions that the provider services. The second tier represents sub-regions, and the third tier represents the records clustered according to the individual clinics corresponding to each sub-region.

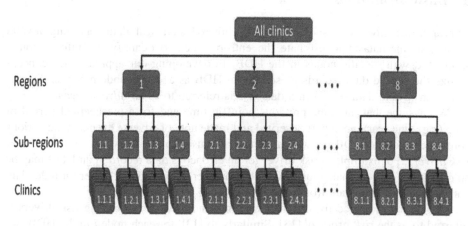

Fig. 1. A schematic structure of a database of pharmaceutical inventory held by a distribution center (adapted from Levner et al. [14])

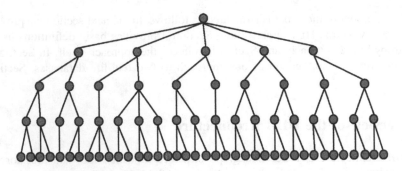

Fig. 2. A structure of a database of car components produced by an automobile manufacturer (adjusted from Levner and Ptuskin [13])

The HDB illustrated in Fig. 2 represents the products produced by a car manufacturer. In this HDB, the root node contains all information about all car components produced by the manufacturer. The first tier contains clusters of components corresponding to large, ready-to-install subsystems, e.g., body assembly, seat assembly, engine module, etc. The second tier represents sub-components of the larger components represented by the first-tier nodes, and so on.

A third example is an HDB containing patient data for a healthcare provider. In this case, the root node depicts all the available records corresponding to patients (see [17, 18]). The nodes (clusters) of the first tier depict clusters of records for the subgroups of patients (e.g., patients of specific age groups). For each of these subgroups, the nodes of the second tier provide the gender division of the information about the patients of different age categories (in the first-tier clusters). The third tier contains, for each of the latter clusters of patient records, sub-clusters of records corresponding to the number of times these patients visited doctors, etc.

3   Basic Definitions

Our approach involves using previously-gathered statistical data regarding invalid records in the dataset to estimate the entropy—i.e., to quantify the information— associated with specific nodes in the HDB. In developing our approach, we conceptualize the invalid data records present in an HDB as a set of random events. Specifically, an invalid data record in a database is referred to as an *adverse event*.

We assume that, at some previous point in time and for a prespecified period of time (e.g., one year), an expert observed and made note of adverse events in the various components of the HDB, as well as potential causes for these events—referred to as *risk drivers*. For example, a risk driver for an erroneous data record in an HDB may be an operator's error or malicious activity by a hacker. A typical risk driver for redundant data is integration of data from multiple sources.

The expert-produced list of adverse events and their corresponding risk drivers is referred to as the *risk protocol* [13]. Similarly to [13], for each node i in the HDB, we assume that the information stored for that node in the risk protocol—i.e., the adverse events that occurred in that node over the prespecified time period and their risk drivers —is recorded in a table, TBL_i. Each row in the table corresponds to an individual adverse event occurring in node i. To the extent that the tables TBL_i, for all the nodes belonging to each HDB tier, are assumed to be given, the statistical data needed to apply our approach can be derived.

Define a tier L_s (also denoted as tier s) as the set of nodes that are at the same distance s from the root node n_0 in the graph of the HDB (see Figs. 1 and 2). Define a *subtree* T_s as a union of the tiers $L_0, L_1, \ldots L_s$.

We assume that, for each node (cluster) of the HDB, the list of *risk drivers* $F = (f_1, f_2, \ldots f_N)$ corresponding to the adverse events in that node is given. For simplicity but without loss of generality, we assume that N, the number of all possible risk drivers corresponding to the node, is the same for all tiers, and that any adverse event is caused by a single risk driver. We call a cluster *invalid* if some invalid records are found in it.

Introduce the following notation:

n_s the total number of clusters in tier s;
$w(s)$ the number of invalid clusters in s registered in the risk protocol;
$m(i, s)$ the number of records in cluster i of s;
$v_{f_i}(i, s)$ an estimate of the percent (frequency) of invalid records caused by driver f in cluster i of s. We assume that this value is either given by experts or calculated by using one of the standard statistical methods of the survey sampling (see, e.g. [19]).

Consider the following event $A_f(i, s)$:

$A_f(i, s) = \{$the risk driver f is a source of an adverse event in cluster i of tier $s\}$,
$$f = 1, \ldots, N, \; s = 1, 2 \ldots$$

Denote the probability of the event $A_f(i, s)$ by $p_f(i, s)$. Assume that the frequency estimate $v_f(i, s)$ of the event $A_f(i, s)$ can serve as an estimation of the probability of the event $A_f(i, s)$, that is:

$$p_f(i, s) = v_f(i, s). \tag{1}$$

We assume, for simplicity, that the considered adverse events are independent.

Now, we turn to determining the probabilities of adverse events in the entire tier.

Let $p_f(s)$ denote the probability of adverse events caused by risk factor f in tier s. We define $p_f(s)$ as a weighted average of the probabilities $p_f(i, s)$ as follows:

$$p_f(s) = [p_f(1, s) \cdot m(1, s) + \ldots + p_f(n_s, s) \cdot m(n_s, s)] / [m(1, s) + \ldots + m(n_s, s)],$$
$$f = 1, \ldots, N.$$

$$\tag{2}$$

Notice that the summation $p_f(s)$ for any f is taken over all the (invalid) clusters of tier s. This value is used for computing the entropy in the next section.

4 Entropy as a Measure of Information Quantity

We now proceed to show how previously-gathered statistical information about invalid data in an HDB can be used to quantify the information loss that is likely to be attributable to such adverse events. The larger the HDB, the larger the loss. Moreover, the larger the size of the cluster in the HDB, the larger the statistical number of mistakes and redundancies, and, hence, the information loss.

We start by formally defining *information entropy*, as follows [9, 10]. Given a set of events $E = \{e_1, \ldots, e_n\}$ with a priori probabilities of event occurrence $P = \{p_1, \ldots, p_n\}, p_i > 0$, such that $p_i + \ldots + p_n = 1$, the entropy function H is defined by

$$H = -K\Sigma_{i=1,\ldots,n} \, p_i \cdot \log p_i, \tag{3}$$

where K is a constant corresponding to a choice of measurement units.

The main idea of the suggested entropy-based approach is that the information entropy estimates the amount of information contained in a flow of adverse events registered in the risk protocol. Thus, the entropy characterizes the uncertainty, or the absence of our knowledge about the available information in the HBD. Recall that the lower the entropy, the more information (knowledge) is available in the HDB.

Using the above relations (1)–(3), we can compute the entropy $H(s)$ of each tier s by applying the probabilities $p_f(s)$ defined by (2), as follows:

$$H(s) = -K(\Sigma_{f=1,\ldots N} \, p_f(s) \cdot \log p_f(s)) \tag{4}$$

Although each tier contains its own set of associated clusters, entropy $H(s)$ at each tier s estimates, in fact, entropy of the entire original HDB. This observation is due to the fact that, for any tier s, $s = 1.2, \ldots$, all the clusters altogether contain, by their construction, the whole set of all the records in the original database, which are "packed" into the corresponding clusters.

Entropy Dynamics. Consider an arbitrary HDB and let the tier number s be $s = 1, 2, \ldots$ Consider a fragment of the HDB containing the tiers from 1 to t only, where t is some integer. Let us call such a structure t-truncated.

We believe that the normalizing coefficient $K = K_s$ should take into account the amount of time spent on cleaning. Let the time for correcting/removing an adverse event caused by factor f in cluster i in tier s be $T_f(i, s)$. Let the time for correcting/removing all the invalid clusters (nodes) in s be T_s and the time for the retrieval of all the records in s (including the corrected records) in the HDB be T_s^{ret}, where $T_s < T^{ret}$.

Assume that

$$K_s = (T_s^{ret} - T_s)/T_s^{ret}. \tag{5}$$

The share of the cleaning time with respect to the entire retrieval time for tier s, $1 \leq s \leq t$, is: T_s/T_s^{ret}. The cleaning time share in the next tier, $s + 1$, is T_{s+1}/T_{s+1}^{ret}.

The key result is as follows. The entropy $H(s)$, defined as a function of the tier number s satisfying (4) and (5), decreases with increasing s.

To prove this fact, we observe first that ratio K_s in (5) decreases with increasing s. Indeed, as we noted in the previous section, the number of invalid records in any cluster is estimated by the standard statistical methods of survey sampling. For simplicity but without loss of generality, we may assume that the average cleaning time is approximately the same for any record in any cluster in any tier. But the number of different clusters quickly grows with the growth of s; therefore, $T_s^{ret} - T_s$ falls with the growth of s. Further, the percent of invalid records is relatively small; in this case, T^{ret} only slightly decreases with the growth of s. Therefore, K_s in (5) decreases when s increases.

The term $\Sigma f\, p_f . \log p_f$, in definition (4) is bounded from above by $H(0) = \log N$, where N is known (i.e., noted in the risk protocol) and bounded from above. As the ratio K_s decreases with increasing s, then entropy $H(s)$ defined by (4) decreases when s increases.

A similar result about the behavior of entropy in hierarchical systems was obtained and confirmed through experiments in [11, 13, 16, 20].

Stopping Rule. Taking into account the entropy dynamics described above, we can identify the minimal number of steps for the cleaning process to be sufficient, assuming that the decision maker sequentially computes the entropy for the HDB in each tier, one after another, starting with the root node. Specifically, such an iterative cleaning process can be stopped as soon as the entropic decrease gradient *grad* becomes sufficiently small upon moving to the next tier. Formally, we define an allowed inaccuracy level, δ, and the iterative cleaning process stops at a *critical tier* s_0 when the following holds:

$$grad = (H(s_0) - H(s_0 + 1))/H(s_0) \leq \delta. \tag{6}$$

This stopping rule determines the optimal (i.e. minimally required) number of tiers in the hierarchical database that need to be examined; it defines a more streamlined representation of the database, in which less informative tiers are eliminated.

As we shall see in the next section, there is no need to clean the entire large-size database; rather, it is sufficient to find the critical tier, s_0, that satisfies (6) and then to perform the cleaning operations for the clusters found at this tier.

5 Computational Application

We have applied the approach described above to two real-life HDB datasets. The first is a dataset from a real automobile HDB, similar to that presented in the example above (Fig. 2) and described in [13]. The initial dataset contained information for about 12,000 car components. We analyzed the entropy dynamics for a subset of the dataset containing 1,500 nodes. We set $\delta = 0.03$ and observed that for $t = 5$ and larger, the value of *grad* is less than $\delta = 0.03$, indicating that it is possible to stop the cleaning process after the fifth tier of the database (see Fig. 3). This stopping rule produces a truncated database structure with 5 tiers, containing 370 nodes at the 5th tier.

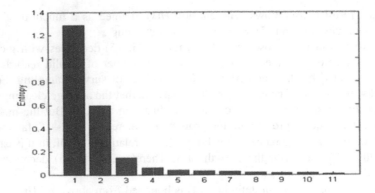

Fig. 3. Results of the entropy computations

The second dataset we used was a massive dataset from a medical data warehouse (DWH) run by Clalit Health Services ("Clalit"), the largest health maintenance organization in Israel [17, 18]. The Clalit DWH comprises historical and current health data for several millions of individuals over the last 25 years. The data include clinical, communicational and administrative data collected from primary care and specialty clinics, hospitals, diagnostic laboratories and medical centers, and national disease registries. The sociodemographic information related to patients and clinics is received periodically from national registries (e.g., Israeli Central Bureau of Statistics).

In prior research [17, 18], we used the Clalit DWH in an investigation focused on patient profile discovery. In this analysis, hundreds of thousands of data records were extracted from the DWH. The authors noticed that a high data granularity allows access to a large number of details regarding each patient but is not efficient for summarizing the data at sub-population levels for supporting the development of new and efficient healthcare policies. Data classification and analysis were done by using the clustering (k-means) method for reducing data complexity. More specifically, at the preprocessing step, the clustering supported data discretization of dozens of administrative and socio-demographical attributes, e.g., location of residence, diagnosis granularity, details of drug treatment, communicational ways to access to the Clalit website, etc. Then the hierarchical clustering during the data mining step helped to discover a very limited number (around a dozen) of patient profiles. Complementary to the studies [17, 18], the suggested entropic approach enabled us to clean the dataset from invalid data and to reduce the dimensionality of the subset extracted from the Clalit DWH.

Herein, in our computational experiments, we applied our entropy-based approach to a subset of DWH data related to patients with diabetes, comprising over 300,000 records as in [18, 19]. However, in contrast to the latter studies, we divided the initial set into other sets of homogenous clusters. We iteratively used the following five attributes for dividing each tier into clusters (categories): (1) age (with 10 categories); (2) gender (2 categories); (3) illness acuteness (3 categories); (4) adherence to treatment (7 categories), and (5) the number of visits of patients to healthcare practitioners (25 categories). We used the stopping rule found in Sect. 4, and set the stopping coefficient at $\delta = 0.02$. The records in each of the clusters of the fourth tier were homogenous with respect to attributes (1)–(4), whereas the records in each of the 5th-tier clusters were homogenous with respect to all the attributes (1)–(5).

The stopping rule indicated that five tiers is sufficient for reaching a stable entropy value in this experiment. Note that in the reduced graph the fourth tier contained about 500 clusters, each one containing about 600 records; and the fifth tier contained about 3,000 clusters each one containing from 90 to 100 records. The behavior of the entropy was observed to be similar to that shown in Fig. 3. As a result of the cleaning and clustering, instead of the original dataset with about 300,000 records, we obtained a reduced graph structure with about 3,000 nodes in the final tier.

6 Concluding Remarks

This paper developed an entropy-based approach for organizing data cleaning in an HDB. This approach enables the decision maker to identify the nodes and tiers in the HDB in which essential quantities of information are likely to have been lost, and to ignore nonessential tiers, thus making the cleaning process more efficient, and producing a reduced representation of the original HDB graph.

We note that combining the data processing (clustering) methodology described in [17, 18] with the entropic approach may enable researchers to streamline their datasets in large-scale investigations (e.g., epidemiological investigations). That is, prior to initiating such an investigation, it could be worthwhile to apply the entropy-based procedure proposed herein, and thus to eliminate redundant information (e.g., obtaining a dozen of the most informative attributes instead of a few hundred). We plan to adopt this approach to further investigate the Clalit DWH, focusing on practical issues of enhancing the influenza vaccination process.

In our future research, we intend to undertake a more in-depth analysis of links between entropy and costs in large-scale databases and to perform cost-benefit-risk analysis of the HDBs, considering the stochastic characteristics of data types.

Acknowledgement. The research of Eugene Levner, Arriel Benis and Shai Ashkenazi is supported by the grant of the Ariel University and Holon Institute of Technology (Israel) "Applications of Artificial Intelligence for enhancing efficiency of vaccination programs" [no. RA19000000649]. The research of Alexander Ptuskin is supported by the grant of the Russian Foundation for Basic Research "Development of economic-mathematical methods for evaluation of alternative options and identification of the best available technologies" [no. 18-410-400001].

References

1. Lee, M.L., Lu, H., Ling, T.W., Ko, Y.T.: Cleansing data for mining and warehousing. In: Bench-Capon, T.J.M., Soda, G., Tjoa, A.M. (eds.) DEXA 1999. LNCS, vol. 1677, pp. 751–760. Springer, Heidelberg (1999). https://doi.org/10.1007/3-540-48309-8_70
2. Rahm, E., Do, H.H.: Data cleaning: problems and current approaches. IEEE Data Eng. Bull. **23**, 1–11 (2000)
3. Volkovs, M, Chiang, F., Szlichta J., Miller, R.J.: Continuous data cleaning. In: 2014 IEEE 30th International Conference on Data Engineering, pp. 244–255 (2014)

4. Khedri, R., Chiang, F., Sabri, K.E.: An algebraic approach towards data cleaning. Procedia Comput. Sci. **21**, 50–59 (2013). https://www.sciencedirect.com/science/article/pii/ S1877050913008028#aep-article-footnote-id2
5. Green, T.J., Tannen, V.: Models for incomplete and probabilistic information. In: Grust, T., et al. (eds.) EDBT 2006. LNCS, vol. 4254, pp. 278–296. Springer, Heidelberg (2006). https://doi.org/10.1007/11896548_24
6. Zimielinski, T., Lipski, W.: Incomplete information in relational databases. In: Readings in Artificial Intelligence and Databases, pp. 342–360 (1989)
7. Shannon, C.E.: A mathematical theory of communication. ACM SIGMOBILE Mob. Comput. Commun. Rev. **5**(1), 3–55 (2001)
8. Stone, J.V.: Information Theory. University of Sheffield, England (2014)
9. Zhou, X., Zhang, Y., Hao, S., Li, S.: A new approach for noise data detection based on cluster and information entropy. In: Proceedings of the 2015 IEEE International Conference on Cyber Technology in Automation, Control, and Intelligent Systems (CYBER), Shenyang, China, pp. 1416–1419 (2015)
10. Kumar, V., Rajendran, G.: Entropy based measurement of text dissimilarity for duplicate-detection. Mod. Appl. Sci. **4**(9), 142–146 (2010)
11. Hebron, A., Levner, E., Hovav, S., Lin, S.: Selection of most informative components in risk mitigation analysis of supply networks: an information-gain approach. Int. J. Innov. Manag. Technol. **3**(3), 267–271 (2012)
12. Allesina, S., Azzi, A., Battini, D., Regattieri, A.: Performance measurement in supply chains: new network analysis and entropic indexes. Int. J. Prod. Res. **48**(8), 2297–2321 (2010)
13. Levner, E., Ptuskin, A.: An entropy-based approach to identifying vulnerable components in a supply chain. Int. J. Prod. Res. **53**(22), 6888–6902 (2015)
14. Hovav, S., Tell, H., Levner, E., Ptuskin, A., Herbon, A.: Healthcare analytics and big data management in influenza vaccination programs: use of information-entropy approach. In: Rodriguez-Taborda, E. (ed.) The Analytics Process: Strategic and Tactical Steps, pp. 211–237. Taylor & Francis, CRC Press, Boca Raton (2017)
15. Isik, F.: An entropy-based approach for measuring complexity in supply chains. Int. J. Prod. Res. **48**(12), 3681–3696 (2010)
16. Karp, A., Ronen, B.: Improving shop floor control: an entropy model approach. Int. J. Prod. Res. **30**(4), 923–938 (1992)
17. Benis, A., Harel, N., Barak Barkan, R., Srulovici, E., Key, C.: Patterns of patients' interactions with a health care organization and their impacts on health duality measurements: protocol for a retrospective cohort study. JMIR Res. Protoc. **7**(11), e10734 (2018)
18. Benis, A.: Identification and description of healthcare customer communication patterns among individuals with diabetes in Clalit Health Services: A retrospective database study for applied epidemiological research. Stud. Health Technol. Inform. **244**, 38–42 (2017)
19. Sapsford, R., Jupp, V.: Data Collection and Analysis. SAGE Publications, London (2006)
20. Kriheli, B., Levner, E.: Entropy-based algorithm for supply-chain complexity assessment. Algorithms **11**(4), 1–15 (2018)

A Performance Prediction Model
for Spark Applications

Muhammad Usama Javaid$^{(\boxtimes)}$, Ahmed Amir Kanoun, Florian Demesmaeker,
Amine Ghrab, and Sabri Skhiri

EURA NOVA R&D, Mont-Saint-Guibert, Belgium
{usama.javaid,amir.kanoun,florian.demesmaeker,
amine.ghrab,sabri.skhiri}@euranova.eu

Abstract. Apache Spark is a popular open-source distributed processing framework that enables efficient processing of massive amounts of
data. It has a large number of configuration parameters that are strongly
related to performance. Spark performance, for a given application, can
significantly vary because of input data type and size, design & implementation of algorithm, computational resources and parameter configuration. So, involvement of all these variables makes performance prediction very difficult. In this paper, we take into account all the variables
and try to learn machine learning based performance prediction model.
We ran extensive experiments on a selected set of Spark applications that
cover the most common workloads to generate a representative dataset
of execution time. In addition, we extracted application and data features to build a machine learning based performance model to predict
Spark applications execution time. The experiments show that boosting
algorithms achieved better results compared to the other algorithms.

1 Introduction

Nowadays it is common for companies to face big data challenges, as they deal
with increasingly large and complex data. Yet, they often struggle to reduce the
processing time of such amount of data. Spark is a popular distributed processing
framework used in big data. However, it comes with too many parameters (above
200 parameters) that need to be tuned to get the best performance. Hence, it
is not easy to manually choose the values of the parameters that will lead to
the best performance. Also, not all the parameters are equally important and
does not have the same impact on performance. So more often in practice, only
a subset of spark parameters is tuned.

The elaboration of this scientific paper was supported by the Ministry of Economy,
Industry, Research, Innovation, IT, Employment and Education of the Region of Wal-
lonia (Belgium), through the funding of the industrial research project Jericho (con-
vention no. 7717).

S. Nepal et al. (Eds.): BIGDATA 2020, LNCS 12402, pp. 13–22, 2020.
https://doi.org/10.1007/978-3-030-59612-5_2

In this paper, we discuss the influence of a subset of Spark parameters on the commonly used applications in industry. We present our work in three main steps. First, a framework to run a set of experiments in Spark and to record aggregate measures of execution time. Second, we discuss the selection of spark parameter and extraction of spark job features to capture the job based effect on execution time. Third, a performance model to predict spark application execution time regarding a set of spark parameters, application features and input data size.

Therefore, through this paper, we aim to build a robust performance model using machine learning algorithms to predict the execution time of a given Spark application. To train a machine learning model, we generated an execution time dataset of various spark applications. Different spark applications were executed with different sets of spark platform parameters along with various sizes of input data and execution time was recorded. The most widely used spark jobs were selected for the generation of the dataset. As shown in [1], some of the existing works tried to predict Spark application performance using execution time as measure. However, the work is limited to a small subset of algorithms and did not explore the application native features.

To overcome this limitation, we built a performance prediction framework that, given a Spark application and the chosen configuration values, extracts the application features and combine them with the dataset to predict the execution time of the application. The contributions of our work can be summarized as follows:

- We built a representative training dataset by identifying the most common Spark application families and running extensive experiments with these families.
- We extracted applications features at different levels (application, stage and job) and input data features to built a performance model using machine learning algorithms to predict Spark applications execution time.
- We built different machine learning prediction models to predict execution time of a Spark application and discussed their accuracy.

The remaining part of this paper is structured as follows: Sect. 2 reviews the state of the art in the domain of performance models for distributed processing frameworks. Section 3 presents overview of the framework and building blocks of the framework. Section 4 discusses the results and Sect. 5 concludes the paper.

2   Related Work

In the recent times, big data frameworks have been very popular as data being produced is increasing rapidly. Many big data frameworks have emerged to process this ocean of data. Among these big data analytics platforms, Apache Spark [2], an open source framework with implicit data parallelism and fault tolerance, is the most popular framework for big data analytics. Due to its efficiency and in-memory computation, Apache Spark is preferred over other frameworks like Hadoop which is slower because of its disk based processing.

Performance prediction for big data analytics platforms is one of the emerging domains that received so much attention in the recent time. The performance of an application is affected by the settings parameters of the framework. There has been a significant number of researches on performance prediction for various big data analytics frameworks, in this paper we are focusing on performance prediction for Apache Spark.

Most of the work in the state of the art focused on MapReduce computing framework or Hadoop platform. Starfish [3] uses a cost-based modeling and simulation to find the desired job configurations for MapReduce workloads. AROMA [4] proposes a two phase machine learning and optimization framework for automated job configuration and resource allocation for heterogeneous clouds. Hadoop's scheduler can degrade the performance in heterogeneous environments and introduced a new scheduler called Longest Approximate Time to End in [5]. In [6], the author analyzed the variant effect of resource consumption for map and reduce. In [7], an automated framework for configuration settings for Hadoop and in [8], the presented KMeans based framework recommends configuration for new Hadoop jobs based on similar past jobs which have performed well.

To the best of our knowledge, there is not a lot of work on the Apache Spark performance prediction. A simulation-driven prediction model has been proposed in [9], in order to predict a Spark job performance. In [10], the authors proposed a support vector regression (SVR) approach for an auto-tuning model and they shown its efficiency in terms of computation. In [11], the authors propose a Spark job performance analysis and prediction tool, by parsing the logs of small scale job and then simulating it for different configuration for performance prediction.

Recently, other machine learning based algorithms have shown significant performance. In [12], the authors selected, what they considered as, the 12 most important Spark parameters and shown a significant performance optimization by tuning them. In [13], the authors proposed an approach to reduce the dimensionality. First, a binary classification model is built to predict the execution time of an Apache Spark application under a given configuration. Then, they used second classification model to define the range of performance improvements.

In [14], the authors proposed a parameter tuning framework for Big data analytics with time constraint. In this work, a Spark application first needs to run at a smaller scale with different parameters - called autotune testbed - capturing the effect of parameters on execution time. Then, an autotune algorithm finds the best parameters configuration for an application within a time constraint. In [15], authors present a machine learning approach to predict execution time of a Spark job. For predicting the execution time, they extract the job level features from Directed Acyclic Graph (DAG) of Spark job and combine it with cluster features naming number of machines, number of cores and RAM size to train a machine learning model. Their work is limited to predicting execution time of SQL queries and Machine Learning workloads.

In this paper we propose to learn a Spark performance function. In our approach we do not need to run the application before estimating its performance. Our algorithm applies prior knowledge learned on a dataset that we tried to make as representative as possible of the reality. In this context, for a given Spark application, configuration, resource and input data we can predict the application performance.

3 Overview

In this section, we present the experimental framework for performance prediction of Spark platform. We explore machine learning algorithms and several aspects of the experimental framework such as feature selection, data collection methods and techniques, training, testing and evaluation. Finally, we evaluate these models w.r.t. accuracy of predicting the execution time (performance) and their sensitivity to various variables such as the Spark configuration parameters, training data size or even the application characteristics.

3.1 An Experimental Framework for Performance Prediction of Spark

A performance model for Spark is required in order to predict the execution time of a Spark job. Here is the Eq. (1) proposed in [16] to denote the performance prediction.

$$perf = F(p, d, r, c) \tag{1}$$

Where *perf* denotes the performance of a Spark application p processing d as input data, running on r resources under the configuration c. As a result, F is the function we want to approximate.

In this paper, our work focuses on performance prediction of a Spark application. Here, the performance is the execution time of a Spark application. In order to learn such a performance model, we needed first to have a dataset representative of the typical Spark applications. Our intuition is that if we are able to create such a dataset, we could end up with an algorithm able to generalise what it learned to predict the performance of any new Spark applications. Therefore, based on our industrial experience in this field, we selected different types of Spark applications: (1) data science workload such as Kmeans, Binomial Logistic Regression, Decision Tree Classifier, Linear Regression, (2) data integration such as SQL Groupby, sorting data, (3) Graph processing such as PageRank, Single Source Shortest Path, Breadth First Search in graphs, (4) general workload such as wordcount. In the rest of the paper, we call these four types of workdloads the four *families*. We ran extensive experiment on these Spark applications and used the approach described in [14] to collect the data. We evaluated two approaches for modeling the performance function. First, we tried to train a dedicated algorithm for each of the four workload family. We had the intuition that a single model would be less accurate in predicting the performance for so many types of workloads. However, in order to valid this first assumption, we tried to train a single model for any kind of applications. We found out that the single model approach was actually way better. In order to create an extensive data set, we ran a significant number of experiments. For a given application p, we ran different input data d, on different resources r with different configurations c. All variables from Eq. 1 have been used in a systematic way to cover all the involved aspects of Spark application affecting the execution time. We divided the result data into four datasets w.r.t. the family it comes from. We evaluated the performance prediction of our algorithm as the ability to correctly predict the execution time.

3.2   Parameter Selection

In this work, we have selected 6 Spark configuration parameters shown in the Table 1, along with the size of the input data and features extracted from execution of Spark applications. The default columns contains the default values for Spark configuration parameters. The search space column defines the range of parameter values considered and explored. These six configuration parameters are selected from [1, 12–14]. In these papers, the authors selected a small subset of spark parameters for parameter tuning. Out of the selected parameters, 4 parameters: spark.shuffle.compress, spark.spill.shuffle.compress, spark.io.compression and spark.reducer.maxSizeInFlight parameters are used in previously mentioned 4 papers. The 2 other parameters naming spark.executor.cores and spark.executor.memory are common in 3 papers and given their direct relevance, made them to make the list for parameters to be considered. The reason to limit to 6 parameters is to cover the effect of important parameters thoroughly. Nevertheless, parameter selection for spark platform not only a different domain

Table 1. Spark parameters and value ranges

Parameter	Default	Search Space	Description
Spark.executor.cores	1	1–8	Number of cores assigned to each executor
Spark.executor.memory	1 GB	1 G-8 G	Size of the memory allocated to each executor
Spark.shuffle.compress	True	True/False	Whether to compress map output files
Spark.shuffle.spill.compress	True	True/False	Whether to compress data spilled during shuffles
Spark.io.compression.codec	lz4	lz4/lzf/snappy	The codec used to compress internal data such as RDD partitions, event log, broadcast variables and shuffle outputs
Spark.reducer.maxSizeInFlight	24 m	24 m–72 m	The max size of map output each reducer could fetch simultaneously

related to Spark platform. This research domain can also be targeted for Spark platform optimization and could possibly lay the foundation for future work.

3.3 Dataset Generation

Ten different Spark workloads naming KMeans, Binomial Logistic Regression, Decision Tree Classifier, Linear Regression, WordCount, Sorting, PageRank, Single Source Shortest Path, Breadth First Search and Groupby were selected to build performance prediction model for Spark. In this paper, we call Spark execution data the result of a spark application run where we collect performance metrics. In order to collect such execution data for different workloads, we adopted an approach inspired by [14]. Experiments are run starting from minimum resources for Spark platform and small input datasets. As the parameter search space is large and could not be explored entirely, we divided it into bins based on domain knowledge and expert opinion. In addition, we also used Bayesian optimization to identify the potential sub-space where to run more experiments. Experiments were scaled up with respect to resources for the Spark and input dataset size. The strategy is to capture the complex relationship between spark parameters and input datasets size for different spark applications. Also, each experiment is run 3 times in order to capture the variance in performance. These experiments were run on Google Kubernetes Engine.

3.4 Application Features Selection

In order to build a robust prediction model, application native features were extracted and included in dataset. Based on our experimentation and state-of-the-art, taking into account the input dataset size only, is not enough. Thus, we aimed at leveraging significant information regarding how Spark deals with a given application and an input dataset. To do so, we rely on the execution graph of a Spark application. This graph is called the Directed Acyclic Graph (DAG). With in a DAG, the nodes represent the Spark internal state (RDD) while the edges represent the actions to transform an RDD to another. Spark application DAG is leveraged because it contains the application complexity per se.

Since we want to build a machine learning model to predict execution time, the DAG needed to be summarized for our model. In fact, in Spark core, Spark Driver is responsible for creating execution plans, also known as Jobs, for any Spark application. Once the plan is created, each job, represented by a DAG, is divided by the driver into a number of stages. These stages are then divided into smaller tasks before sending them to workers (executors) in order to be executed. Spark prepares the logical plan on the Spark Driver before the execution step, this plan defines the steps and operation to be performed for the execution of an application. Fortunately, Spark platform has its history Server's web UI from which the following features were extracted:

- Number of jobs
- Aggregate
- Collect
- Count
- Number of completed tasks
- Number of completed stages

Execution Time as a Performance Measure. Several measures can be taken into consideration for a batch parallel processing job like speedup, efficiency and redundancy. However, those performance measures can be related to a benchmark between one processing unit and multiple processing units or comparing sequential execution to parallel execution. In this work, we focus on recording execution time as a performance measure for the experiments we run.

3.5   Performance Model

The purpose of this work as mentioned above is to predict the execution time of a given Spark application under specified spark configuration. In this section, we present the performance model used to make this prediction.

The dataset is composed of 3 kinds of features: Spark parameters, application features and input data size as a dataset feature in addition to application execution time as target. Each row of the data represents an experiment run on a cluster with given Spark configuration, a Spark workload and a dataset. Given the complex relationship and vast range of spark parameters adds to the complexity of the problem, making it harder to predict the execution time with accuracy.

Different machine learning regressor algorithms: Linear Regression, Boosting, Random Forest and Neural Networks were implemented to build a performance model.

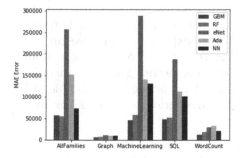

Fig. 1. Accuracy MAE application features ingested

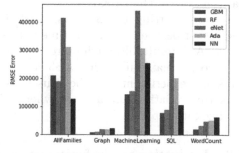

Fig. 2. Accuracy RMSE application features ingested

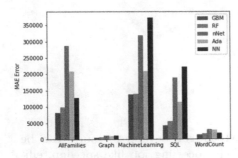

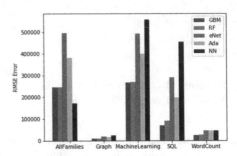

Fig. 3. Accuracy MAE application features not ingested

Fig. 4. Accuracy RMSE application features not ingested

4   Experimental Evaluation

4.1   Experimental Setup

In this work, each experiment is an execution of a Spark application with defined set of Spark parameters values, application features and dataset feature (size). We proceed in following steps: First, all selected spark applications were run with all the possible combination of selected spark parameters. Each experiment was run 3 times in a Kubernetes cluster. Spark application features are extracted for each application. Finally, all the experiments results are appended to create a dataset for machine learning algorithms.

We used Google Kubernetes Engine with 5 nodes with three different resource configuration according to the size of the input dataset. The three different resource configurations are 1 core & 1 GB RAM per node, 4 cores & 4 GB RAM per node and 8 cores & 8 GB RAM per node.

Each application was run with an input datasets of different sizes. We classified the datasets regarding the size into 3 major classes: (1) small: for datasets less than 1GB, (2) medium: for datasets between 1GB and 10 GB, (3) big: for datasets bigger than 10GB.

Before starting the modeling part, it was important to check a potential dispersion between execution time values for one experiment since we run it 3 times. As a result, although we record a 6.73% as coefficient of variation's mean, we record a maximum value of 83.06% for the same metric. This large variation for some experiments can be caused by networking latency for example. We took this variation into account when building the performance model in order to check its impact.

4.2   Evaluation of Performance Model's Accuracy

The machine learning prediction model was built under two different settings. First model was built without taking into account the application features and the later was built including the application features. The reason for this approach was to check the contribution of application features in the prediction of execution time by the models. Figure 1, Fig. 2 show both mean absolute error (MAE) and root mean squared error (RMSE) of the machine learning algorithms per family with application features ingested and Fig. 3, Fig. 4 show both mean absolute error (MAE) and root mean squared error (RMSE) of the machine learning algorithms per family without application features ingested.

On average, Gradient Boosting Machine (GBM) and Random Forest (RF) perform the best compared to other methods. The execution time's mean is 399025 ms, the mean absolute error of GBM is 46662 and for Random Forest it was 44621. The error is approximately 10% for all families together. All ML algorithms provide good results on Word Count and Graph family thanks to the instrinsic simplicity of the chosen workloads. However, it proved difficult to predict the execution time of the Machine Learning family because of different types and complexity of the workloads, as well as for the GroupBy in SQL.

The performance model was almost the same with and without the application, however for some families of workloads, slight improvement was recorded.

5   Conclusion

In this paper, we presented a novel method for predicting the performance of a Spark application using machine learning models. An experimental framework for generating relevant datasets for performance prediction was developed and several machine learning algorithms were evaluated. Experimental results show that GBM and RF have a good accuracy and computational performance across diverse spark workloads.

In the future, further research could be conducted on parameter selection in order to add more suitable spark parameters. The other revelent aspects affecting the performance prediction, such as the resource competition between the jobs and fluctuation of network bandwidth could also be explored.

References

1. Chao, Z., Shi, S., Gao, H., Luo, J., Wang, H.: A gray-box performance model for apache spark. Future Gener. Comput. Syst. **89**, 58–67 (2018)
2. Zaharia, M., et al.: Resilient distributed datasets: a fault-tolerant abstraction for in-memory cluster computing. In: Proceedings of the 9th USENIX Conference on Networked Systems Design and Implementation, NSDI 2012, Berkeley, CA, USA, p. 2. USENIX Association (2012)
3. Herodotou, H., et al.: Starfish: a self-tuning system for big data analytics. In: CIDR, pp. 261–272 (2011)
4. Lama, P., Zhou, X.: Aroma: automated resource allocation and configuration of MapReduce environment in the cloud. In: Proceedings of the 9th International Conference on Autonomic Computing, ICAC 2012, pp. 63–72. ACM, New York (2012)
5. Zaharia, M., Konwinski, A., Joseph, A.D., Katz, R., Stoica, I.: Improving MapReduce performance in heterogeneous environments. In: Proceedings of the 8th USENIX Conference on Operating Systems Design and Implementation, OSDI 2008, Berkeley, CA, USA, pp. 29–42. USENIX Association (2008)

6. Kambatla, K., Pathak, A., Pucha, H.: Towards optimizing Hadoop provisioning in the cloud. Hot Top. Cloud Comput. **9**, 28–30 (2009)

7. Wu, D., Gokhale, A.: A self-tuning system based on application profiling and performance analysis for optimizing Hadoop MapReduce cluster configuration. In: 20th Annual International Conference on High Performance Computing, pp. 89–98, December 2013

8. Zhang, R., Li, M., Hildebrand, D.: Finding the big data sweet spot: towards automatically recommending configurations for Hadoop clusters on docker containers. In: 2015 IEEE International Conference on Cloud Engineering, pp. 365–368, March 2015

9. Wang, K., Khan, M.M.H.: Performance prediction for apache spark platform. In: Proceedings of the 2015 IEEE 17th International Conference on High Performance Computing and Communications, 2015 IEEE 7th International Symposium on Cyberspace Safety and Security, and 2015 IEEE 12th International Conference on Embedded Software and Systems, HPCC-CSS-ICESS 2015, Washington, DC, USA, pp. 166–173. IEEE Computer Society (2015)

10. Yigitbasi, N., Willke, T.L., Liao, G., Epema, D.H.J.: Towards machine learning-based auto-tuning of MapReduce. In: 2013 IEEE 21st International Symposium on Modelling, Analysis and Simulation of Computer and Telecommunication Systems, pp. 11–20 (2013)

11. Singhal, R., Phalak, C., Singh, P.: Spark job performance analysis and prediction tool, pp. 49–50, April 2018

12. Gounaris, A., Torres, J.: A methodology for spark parameter tuning. Big Data Res. **11**, 22–32 (2017)

13. Wang, G., Xu, J., He, B.: A novel method for tuning configuration parameters of spark based on machine learning. In: 2016 IEEE 18th International Conference on High Performance Computing and Communications, IEEE 14th International Conference on Smart City, IEEE 2nd International Conference on Data Science and Systems (HPCC/SmartCity/DSS), pp. 586–593, December 2016

14. Bao, L., Liu, X., Chen, W.: Learning-based automatic parameter tuning for big data analytics frameworks. CoRR, abs/1808.06008 (2018)

15. Mustafa, S., Elghandour, I., Ismail, M.A.: A machine learning approach for predicting execution time of spark jobs. Alexandria Eng. J. **57**(4), 3767–3778 (2018)

16. Herodotou, H., Babu, S.: Profiling, what-if analysis, and cost-based optimization of MapReduce programs (2011)

Predicting the DJIA with News Headlines and Historic Data Using Hybrid Genetic Algorithm/Support Vector Regression and BERT

Benjamin Warner, Aaron Crook, and Renzhi Cao[✉]

Pacific Lutheran University, Tacoma, WA 98447, USA
b.c.warner@wustl.edu, {crookaj,caora}@plu.edu

Abstract. One important application of Artificial Intelligence (AI) is forecasting stock price in the stock market, as such knowledge is highly useful for investors. We first examine two state-of-the-art AI techniques, hybrid genetic algorithm/support vector regression and bidirectional encoder representations from transformers (BERT). After that, we proposed a new AI model that uses hybrid genetic algorithm/support vector regression and BERT to predict daily closes in the Dow Jones Industrial Average. We found that there is an up to 36.5% performance improvement with root-mean-squared-errors using headline data compared to without headline data based on models we have tested, although further analysis may reveal more significant improvements. The code and data used in our model can be found at: https://github.com/bcwarner/djia-gasvr-bert.

1 Introduction

Artificial Intelligence has been widely used in different fields and achieved great performance, including image recognition, language translation, protein structure prediction, finance, *etc.* [1–7]. In the finance applications, there are several types of closely-related financial problems where artificial intelligence can be applied, such as stock market forecasting, portfolio generation, central bank forecasts, and so on. In this paper, we primarily examine *stock market forecasting*, as knowing the direction the market may go—*bearish* or *bullish*—can allow us to make more sound financial decisions. For example, if we predict with confidence that the DJIA will decline tomorrow, and we have a portfolio that is primarily composed of DJIA stocks, we might sell them in the prospect of buying them back at a discounted price the next day. Conversely, if our portfolio had a lot of foreign stocks, and a number of our foreign indices, like the Nikkei 225, indicating that the market would be on the upswing the next day, we might buy more foreign stocks in the hopes of seeing a return the next day. For this type of problem, we could attempt to predict the direction of the overall market (as described with an index like the S&P 500), with data like moving averages, previous stock market data, news headlines, *etc.*

© Springer Nature Switzerland AG 2020
S. Nepal et al. (Eds.): BIGDATA 2020, LNCS 12402, pp. 23–37, 2020.
https://doi.org/10.1007/978-3-030-59612-5_3

Stock market forecasting has existed long before artificial intelligence techniques became useful. Writing in 1973, Graham wrote that "[for] years the financial services have been making stock-market forecasts without anyone taking this activity very seriously. Like everyone else in the field they are sometimes right and sometimes wrong" [8]. Instead of forecasting, most investing in the 1970s followed the school of *fundamental analysis*, which attempts to measure the value of stocks through "a careful analysis of the relevant facts" [9]. Decisions would have been based upon both qualitative and quantitative factors, and would have avoided speculative behavior. *Quantitative finance* would begin to take prominence in the 1990s, when it was realized that future returns could be predicted with accuracy on the basis of accounting and technical measures [9]. With artificial intelligence techniques, which effectively extends from quantitative finance, we can now forecast the market with higher accuracy than many forecasters of the fundamental analysis era.

Among the many artificial intelligence techniques for stock market forecasts that we shall examine are *hybrid genetic algorithm/support vector regression* and *bidirectional encoder representations from transformers* (BERT). From there we will examine a combination of hybrid genetic algorithm/support vector regression and BERT, trained upon DJIA closing prices and article headlines from the New York Times. Finally, we shall examine the efficacy of GA/SVR, SVR, and kernel ridge regression when headline data is used, with results ranging from a 29.2% increase in RMSE to a 36.5% decrease in RMSE.

2   Literature Review

2.1   Hybrid Genetic Algorithm/Support Vector Regression

One important stock forecasting technique is *hybrid genetic algorithm/support vector regression*. In this technique, we apply *multivariate adaptive regression splines*, *kernel ridge regression*, and *stepwise regression* to find the best technical indicators for a model to be built off of. Then we evaluate potential solutions that are built off of the *support vector regression* model using the *genetic algorithm*. Support vector regression is used in contrast to some of the other forms of regression as SVR will not generate certain hypotheses outside of a certain tolerance [10]. SVR will, however, evaluate examples outside of the tolerance range with a penalty [11]. The genetic algorithm is then used to evaluate the regression models by generating new *offspring* based off of a candidate solution, repeatedly selecting a solution until either 2000 iterations (*generations*) have been made, or the solution came within a certain root-mean-squared error. The authors tested several models on a number of Taiwanese stocks, many of which did not involve the genetic algorithm, and found that MARS/GA/SVR and SR/GA/SVR tended to outperform the non-GA models. The root-mean-squared errors for these tended to range between ~ 0.0087 and ~ 0.7413, which were significantly better than their non-GA counterparts, which ranged from ~ 0.180 to ~ 125.99 [10].

2.2 Bidirectional Encoder Representations from Transformers

Bidirectional Encoder Representations from Transformers, or BERT, is a new classification technique from Google primarily used with text inputs for AI [13]. Its unique advantage over other texts analyzers is that it does not read directly left-to-right or right-to-left, rather it considers both directions, using an attention model called a transformer [14], enabling it to decipher the context of a word based on all of the other words in a sentence. The tokens are embedded into vectors and then processed in the neural network. The training of a BERT program involves taking a sentence, masking 15% of the words, trying to predict them, as well as guessing the chronological order of sentences [15]. Once trained, a BERT program can guess the intent and context of text input [16]. A group from Hong Kong was able to find relevant results from reputable investor's sentiments via their posts on Weibo, a Chinese social media site [17]. These results, when factored in as an additional variable into an already existing stock market predictor improved the accuracy of the program. BERT represents a great innovation in the field of text-based analysis AI, an area where AI had previously been lacking. The utilization of BERT in the study we reviewed was able not only to decipher meaning and context from countless Weibo posts, but also evaluate the credibility of the author of those posts.

2.3 Proposed Model

We will be examining a combination of GA/SVR and BERT. BERT will be trained upon headlines from the *New York Times* from this same time period—using the NYT's API [18]—particularly from articles relating to the DJIA and component companies, and it will generate a *sentiment index*. We will use the genetic algorithm and support vector regression trained upon DJIA data from within the last 10 years [19] with an added time-lag, as well as the sentiment index from BERT, as adding fundamental and quantitative factors will make it more accurate [20]. GA/SVR will be able to predict accurately in a more quantitative approach, while BERT will help ensure that a fundamental approach is also used, as it relies on textual data. Headlines and DJIA data from the first 80% of the time period will be used to train the model, and the remaining 20% will be test data. We will not be using MARS, KRR, or SR to determine what technical indicators to use as in Tsai, Cheng, Tsai, & Shiu [10], and will instead rely completely on the value of the index. In future, we may consider quantifying our confidence level in a given prediction, which could represent how much stock one should be willing to buy [21].

3 Methods

We collected our headlines from the New York Times' Article Search API[18]. Headlines, lead paragraphs, and abstracts from the API for each given search term and date were collected for the date range of 11/09/2018 to 11/07/2019.

Search terms were chosen semi-arbitrarily as to minimize ambiguity and maximize the number of articles returned, and were either full company names as suggested by the `nytimes.com` search algorithm or were shorthand names (*e.g..* IBM). In one case, with The Travelers Companies, the stock ticker (TRV) was used as it was the least ambiguous. Collection was done by querying one given search term on one given date. The companies that make up the DJIA change over time, and as such, we only collected headline data for companies that were members on a given day. A full list of the member companies and search terms used can be seen in Table 1. Originally, we attempted to collect headline data for the entirety of the span of the FRED DJIA dataset (11/09/2009 to 11/08/2019), but resource limitations restricted us to access the last year of the data set (11/09/2018 to 11/07/2019). After collecting these headlines, lead paragraphs, and abstracts, we used BERT's sentiment analysis function to classify articles as positive or negative. We ran the BERT model using Tensorflow 1.15 on a Nvidia Tesla K80 using a Google Cloud Compute Engine VM.[1] After this, we analyzed the number of positive and negative examples for a given day and produced two measures, the positive/negative/total quantity for a given day and the *net positivity ratio*, which are discussed in more detail in the Results section. Then we regressed upon headlines for the previous day and the closing prices for the previous two days to generate the prediction for a given day. We accomplished this using sklearn's support vector regression model[22] guided by Calzolari's genetic algorithm for sklearn[23]. Before regressing, we used sklearn's `MinMaxScaler` to scale the data within the range of $[0,1]$[22]. For regression upon dates, we used Unix timestamps—which are the number of seconds since Janauary 1st, 1970—at midnight on a given day. For comparison, we used `sklearn`'s kernel ridge regression implementation [22].

Initially, we did not filter out irrelevant results collected from the `nytimes.com` API as we did not collect article bodies, and thus could not be certain about correctly filtering out relevant articles. After examining the initial results, we decided to apply a filter to determine if a performance gain could be achieved. We collected lead paragraphs, and filtered out article headlines based upon the following heuristic:

1. Find the original article result from which a headline was derived.
2. If the company name–column 1 of Table 1–is in the headline or the lead paragraph, then keep this headline.
3. If the regular expression `economy|market|Dow Jones|DJIA|bear|bull|debt|earnings|quarter|shares|shareholder|dividend` matches the headline or the lead paragraph, then keep this headline.
4. Otherwise, discard this headline (Fig. 1).

There are a number of limitations to this approach. First, the classification of articles by BERT was trained using a dataset involving movie reviews, which will not necessarily be as accurate as a dataset of news headlines. The second limitation is that an apparently significant number of the articles returned by the

---

[1] This was supported by an education grant from Google Cloud.

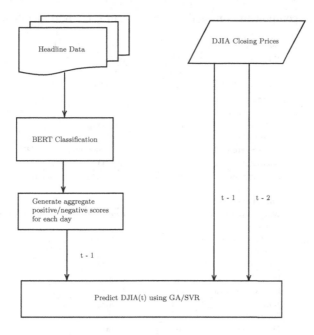

Fig. 1. Flowchart of proposed model. Headline data and DJIA closing prices form the basis of the prediction of the next day's DJIA closing price

New York Times were irrelevant to the desired search terms. Generic keywords like "Inc." or "Visa" lend itself to ambiguity, which is a likely cause of this issue. The third issue is that even when the correct topic is found, relevancy can vary highly, as the importance and degree of relation to the topic searched for will not be reflected in the labelling as negative or positive. A number of resource limitations prevented us from exploring a more robust dataset, and as such further research utilizing a refined and extended dataset may be worthwhile. In addition, there are several limitations of the filtering heuristic discussed above, including that step 3 is not a comprehensive list of financial terms, and it may capture articles that have no direct correlation to any of the components of the Dow Jones. However, this heuristic has demonstrated a noticeable improvement in performance in most models as seen in Table 2, which shall be discussed further in Results.

For the headline data, we regressed upon five versions of the headline data. The first version—which can be seen in Fig. 7—regressed upon the total number of articles for a given day, the number of positive articles in a given day, and the number of negative articles in a given day. The second version of this involved calculating a score that shall be referred to as the *net positivity ratio* (Fig. 4). Given p positive articles and n negative articles, the net positivity ratio is then $\frac{p-n}{p+n}$. A cube root of the net positivity ratio was also tested, but no effect was observed on the RMSE. We regressed upon this score and the total number of articles to generate Fig. 8. We also examined SVR and GA/SVR without

Table 1. DJIA component companies, search term used, and membership dates

Company [24]	Search Term	End	Start	Replacing
3M	3M Company	(end)	(start)	
American Express	AXP	(end)	(start)	
Apple	Apple Inc	(end)	3/19/2015	AT& T [25]
Boeing	Boeing Company	(end)	(start)	
Caterpillar	Caterpillar Inc	(end)	(start)	
Chevron	Chevron Corporation	(end)	(start)	
Cisco Systems	Cisco Systems Inc	(end)	6/8/2009 [29]	N/A [27]
Coca-Cola	Coca-Cola	(end)	(start)	
Dow	Dow Chemical	(end)	3/20/2019 [26]	DowDuPont
ExxonMobil	ExxonMobil Corporation	(end)	(start)	
Walgreens Boots Alliance	Walgreens Boots Alliance	(end)	6/26/2018	General Electric [27]
Goldman Sachs Group	Goldman Sachs Group	(end)	9/23/2013	Bank of America/Altria Group [28]
Home Depot	Home Depot Inc	(end)	(start)	
IBM	IBM	(end)	(start)	
Intel	Intel Corporation	(end)	(start)	
Johnson & Johnson	Johnson & Johnson	(end)	(start)	
JPMorgan Chase	JPMorgan	(end)	(start)	
McDonald's	McDonald's	(end)	(start)	
Merck	Merck & Company	(end)	(start)	
Microsoft	Microsoft	(end)	(start)	
Nike	Nike Inc	(end)	9/23/2013	Alcoa [28]
Pfizer	Pfizer Inc	(end)	(start)	
Procter & Gamble	Procter & Gamble	(end)	(start)	
The Travelers Companies	TRV	(end)	6/1/2009	Citigroup [29]
UnitedHealth Group Inc	UnitedHealth Group	(end)	9/24/2012	Kraft [30]
United Technologies	United Technologies	(end)	(start)	
Verizon	Verizon	(end)	(start)	
Visa	Visa Inc	(end)	9/23/2013	Hewlett-Packard [28]
Walmart	Walmart	(end)	(start)	
Walt Disney	Disney	(end)	(start)	
AT& T	AT& T	3/19/2015 [25]	(start)	
DowDuPont	DowDuPont	3/20/2019 [26]	9/1/2017	DuPont [31]
General Electric	General Electric Company	6/26/2018 [27]	(start)	
Bank of America	Bank of America Corporation	9/23/2013 [28]	(start)	
Alcoa	Alcoa	9/23/2013 [28]	(start)	
Citigroup	Citigroup	6/1/2009 [29]	(start)	
Kraft	Kraft	9/24/2012 [30]	(start)	
Hewlett-Packard	Hewlett-Packard	9/23/2013 [28]	(start)	
DuPont	DuPont	9/1/2017 [31]	(start)	

Table 2. Comparison of different regressions upon headlines

Method	GA/SVR RMSE	SVR RMSE	KRR
Total, Positive, and Negative Quantities	0.036468	0.046145	0.034987
Total Quantity, Net Positivity Ratio	0.036468	0.048525	0.036391
Total, Positive, and Negative Quantities, Filtered	**0.035638**	**0.032663**	**0.034632**
Total Quantity, Net Positivity Ratio, Filtered	**0.035638**	0.046204	0.034803
Total Quantity, Net Cumulative Sum, Filtered	0.036468	0.066482	0.037348
Control	0.036468	0.051466	0.035929

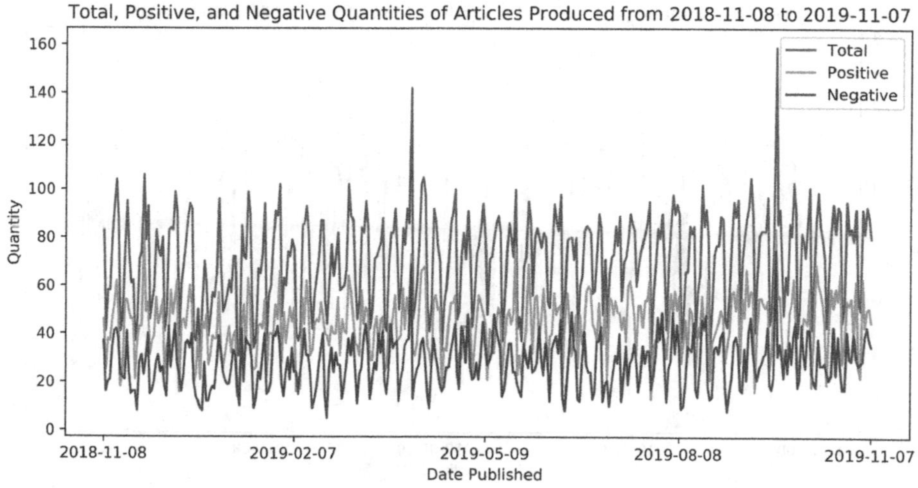

Fig. 2. Total, positive, and negative quantities of articles collected for the desired date range.

regressing upon the headline data (Fig. 12). After examining the results, we then applied the filter and generated new total numbers of articles (Fig. 3) and new net positivity ratios (Fig. 5). We also found the cumulative sum of the filtered positive and negative articles through a given date range—which can be seen in Fig. 6—the difference of which shall be referred to as the *net cumulative sum*. We regressed upon the net cumulative sum and total quantity for that day to generate the results in Fig. 11. It should be noted that this filtered dataset does

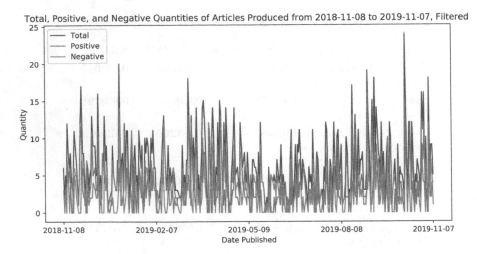

Fig. 3. Total, positive, and negative quantities of articles collected for the desired date range with filter applied.

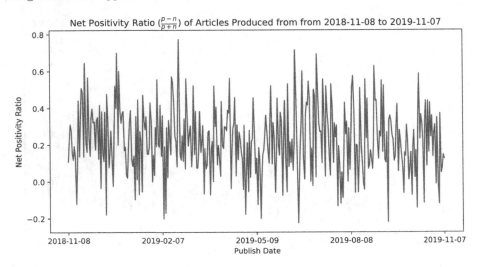

Fig. 4. Net positivity ratio $\frac{p-n}{n+p}$ for the desired date range.

not include data for November 23rd and January 23rd, as our filter eliminated news for those days.

Our model and the data used can be found at https://github.com/bcwarner/djia-gasvr-bert. Instructions on usage can be found in the repository's README.md.

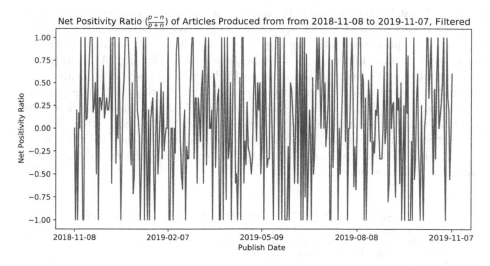

Fig. 5. Net positivity ratio $\frac{p-n}{n+p}$ for the desired date range with filter applied.

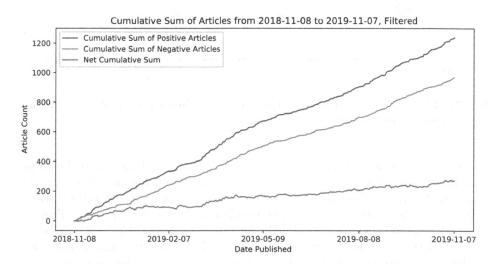

Fig. 6. Cumulative sum of positive, negative, and net articles for the desired date range with filter applied.

4   Results

We found that for the articles collected without filtering, no discernible pattern could be observed for either the quantities of positive/negative articles in Fig. 2 or in the net positivity ratio for Fig. 4. It is highly cyclical, which appears to be a result of the difference in weekday/weekend publishing. A limitation in this is that any news that is reported on Friday will be given no consideration, unless the reporting for a given topic continues on Sunday. With the filtering, total quantities (Fig. 3) and net positivity (Fig. 5) again have no discernible pattern, although both measures tended towards extremes more often. The net cumulative sum, shown in Fig. 6, shows a discernible tendency for there to be more positive articles related to the Dow Jones, which reflects an overall bull market during this time period.

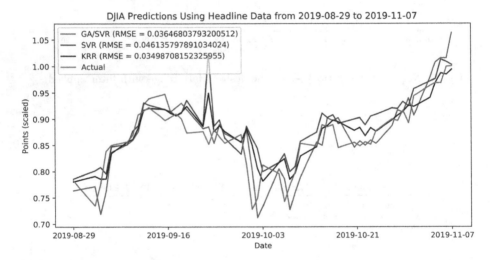

Fig. 7. GA/SVR, SVR, KRR with Total, Positive, Negative Quantities vs. Actual DJIA Closing Prices (scaled)

As can be seen in Table 2, when using GA/SVR, both unfiltered headline measures and the control produce equal RMSEs. When using only SVR, we find that headline quantities tend to perform better than net positivity ratio, both of which performed better than using no headline data, which is visible in Table 2. The lack of change between the RMSEs for the variations of GA/SVR may be an indication that GA gave minimal weight to the headline data, although further analysis is required to confirm this. In addition, we can see that in many cases, the predicted value of the DJIA tends to follow the direction that it went a day afterwards. Interestingly, kernel ridge regression tended to outperform GA/SVR by a slight margin, and SVR by a noticeable margin in terms of RMSE. After applying a filter, we see a minor performance improvement for GA/SVR in total

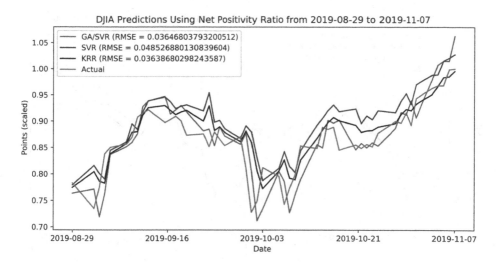

Fig. 8. GA/SVR, SVR, KRR with Total Quantity, Net Positivity Ratio vs. Actual DJIA Closing Prices (scaled)

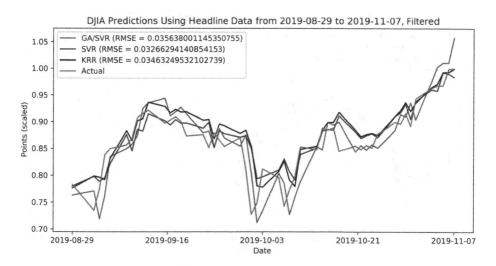

Fig. 9. GA/SVR, SVR, KRR with Total, Positive, Negative Quantities vs. Actual DJIA Closing Prices with Filter Applied (scaled)

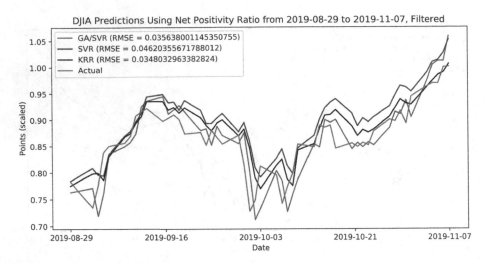

Fig. 10. GA/SVR, SVR, KRR with Total Quantity, Net Positivity Ratio vs. Actual DJIA Closing Prices with Filter Applied (scaled)

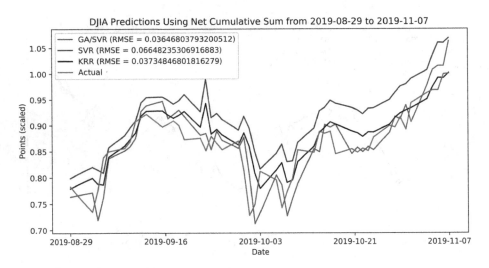

Fig. 11. GA/SVR, SVR, KRR with Total Quantity, Net Cumulative Sum vs. Actual DJIA Closing Prices with Filter Applied (scaled)

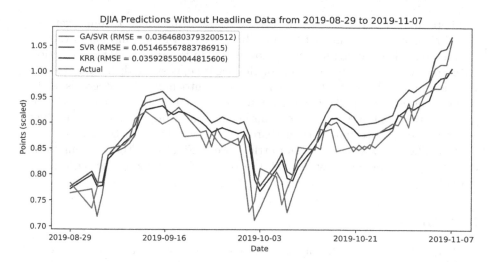

Fig. 12. GA/SVR, SVR, KRR with No Headlines vs. Actual DJIA Closing Prices (scaled)

quantities and net positivity ratio, as well as improvements in SVR and KRR, visible in Fig. 9. For total quantity/net cumulative sum (Fig. 11), we find that GA/SVR performs no differently than the control, SVR performs poorly, and KRR performs slightly worse. Overall, SVR regressed upon total, positive, and negative quantities performed the best in predicting the DJIA (Fig. 10).

5 Discussion

We found that for GA/SVR, using headlines provided no significant improvement in terms of making accurate predictions. Without using the genetic algorithm, SVR tended to perform better with the headline metrics used. In addition, it appears that, while low RMSEs can be achieved, the direction of these predictions tend to follow rather than precede the real closing values. This implies that using the previous two closing values is not enough to make a useful prediction. Moreover, it may imply that headlines may not be enough to predict the DJIA, and delving into more fundamental analysis—or further into quantitative analysis—may be necessary to achieve reliable predictions.

For BERT, the model we trained was able to categorize headlines according to their positive or negative sentiment, but this did not always directly correspond to the values displayed in the DJIA. Journalistic practices with regards to headline writing as well as missing contextual information have the potential to decrease the accuracy of results. The way we used BERT resulted only in marginal improvements on certain models. Based on our results, BERT shows great potential with further refinement of the model and dataset used, as it incorporates a more fundamental approach.

There are several areas where further research may lead to improvements. First, improving the quality of the headlines selected for evaluation in terms of relevancy may lead to significant improvements in terms of BERT's predictions. Switching from classifying in terms of the negative/positive dichotomy to using fuzzy logic—which allows for an entity to fit into multiple categories with a *membership function* [12]—may also refine our usage of BERT, as not all positive/negative events are the same. Another potential improvement would be to weight the articles to match their search term's weighting in the DJIA, as the DJIA is not equally weighted as our articles were. Finally, further examination of other regression models may be useful in terms of maximizing the performance of predictions.

References

1. Hou, J., Wu, T., Cao, R., Cheng, J.: Protein tertiary structure modeling driven by deep learning and contact distance prediction in CASP13. Proteins: Struct. Funct. Bioinform. **87**(12), 1165–1178 (2019)
2. Lecun, Y., Bengio, Y., Hinton, G.: Deep learning. Nature **521**, 436 (2015)
3. Cao, R., Freitas, C., Chan, L., Sun, M., Jiang, H., Chen, Z.: ProLanGO: protein function prediction using neural machine translation based on a recurrent neural network. Molecules **22**, 1732 (2017)
4. Deng, J., Dong, W., Socher, R., Li, L., Li, K., Fei-Fei, L.: ImageNet: a large-scale hierarchical image database (2009)
5. Stephenson, N., et al.: Survey of machine learning techniques in drug discovery. Curr. Drug Metab. **20**, 185–193 (2019)
6. Conover, M., Staples, M., Si, D., Sun, M., Cao, R.: AngularQA: protein model quality assessment with LSTM networks. Comput. Math. Biophys. **7**, 1–9 (2019)
7. Kosylo, N., et al.: Artificial intelligence on job-hopping forecasting: AI on job-hopping (2018)
8. Graham, B.: The Intelligent Investor (2009)
9. Sloan, R.G.: Fundamental analysis redux. SSRN Electron. J. (2018). https://doi.org/10.2139/ssrn.3176340
10. Tsai, M.-C., Cheng, C.-H., Tsai, M.-I., Shiu, H.-Y.: Forecasting leading industry stock prices based on a hybrid time-series forecast model. PLoS One **13**, 1–24 (2018)
11. Smola, A.J., Schölkopf, B.: A tutorial on support vector regression. Stat. Comput. **14**, 199–222 (2004). https://doi.org/10.1023/B:STCO.0000035301.49549.88
12. Atsalakis, G.S., Dimitrakakis, E.M., Zopounidis, C.D.: Elliott Wave Theory and neuro-fuzzy systems, in stock market prediction: the WASP system. Expert Syst. Appl. **38**, 9196–9206 (2011)
13. Devlin, J., Chang, M.-W., Lee, K., Toutanova, K.: BERT: pre-training of deep bidirectional transformers for language understanding. Google AI Language (2019). arXiv:1810.04805v2
14. Vaswani, A., et al.: Attention is All You Need (2017). arXiv:1706.03762v5
15. Rani, H.: BERT - State of the art Language Model for NLP. LyrnAI (2019)
16. Shuyi, W.: Text Classification with BERT and Tensorflow in Ten Lines of Code. Towards Data Science (2019)
17. Hiew, J., et al.: BERT-based financial sentiment index and LSTM-based stock return predictability. arXiv (2019). arXiv:1906.09024v1

18. New York Times. Article Search. New York Times Developer
19. Federal Reserve Bank of St. Louis. Dow Jones Industrial Average (DJIA)
20. Zheng, A., Jin, J.: Using AI to Make Predictions on Stock Market (2017)
21. Das, S., Sen, R.: Sparse portfolio selection via bayesian multiple testing. arXiv (2019). arXiv:1705.01407v3
22. Pedregosa, F., et al.: Scikit-learn: machine learning in Python. J. Mach. Learn. Res. **12**, 2825–2830 (2011)
23. manuel-calzolari/sklearn-genetic: sklearn-genetic 0.2. https://doi.org/10.5281/zenodo.3348077
24. S & P Dow Jones Indices. The Changing DJIA. Indexology. https://us.spindices.com/indexology/djia-and-sp-500/the-changing-djia. Accessed 21 Nov 2019
25. Park, J.Y.: Apple to replace AT&T in the Dow Jones Industrial Average. CNBC (2015)
26. NYSE. DOW INC DOW. NYSE. https://www.nyse.com/quote/XNYS:DOW. Accessed 21 Nov 2019
27. LaVito, A.: GE booted from the Dow, to be replaced by Walgreens. CNBC (2018). https://www.cnbc.com/2018/06/19/walgreens-replacing-ge-on-the-dow.html. Accessed 21 Nov 2019
28. Brooks, R., Waggoner, J., Krantz, M.: Dow 30 adds Goldman Sachs, Nike and Visa. USA Today (2013). https://www.usatoday.com/story/money/markets/2013/09/10/changes-dow-jones-industrial-average/2791723/. Accessed 21 Nov 2019
29. Browning, E.S.: Travelers, Cisco replace Citi, GM in Dow. Wall Street J. (2009). https://www.wsj.com/articles/SB124386244318072033. Accessed 21 Nov 2019
30. Haar, D.: UnitedHealth Group Joins Dow Jones Index, Giving Hartford A Lift. Hartford Courant (2018). https://www.courant.com/business/hc-xpm-2012-09-24-hc-haar-unitedhealth-dow-jones-20120924-story.html. Accessed 21 Nov 2019
31. Lombardo, C.: Questions remain as Dow and DuPont become DowDuPont. Wall Street J. (2017). Available at: https://www.wsj.com/articles/questions-remain-as-dow-and-dupont-become-dowdupont-1504269341. Accessed 21 Nov 2019
32. Predicting Movie Reviews with BERT on TF Hub.ipynb (2019). https://colab.research.google.com/github/google-research/bert/blob/master/predicting_movie_reviews_with_bert_on_tf_hub.ipynb#scrollTo=-thbodgih_VJ. Accessed 24 Nov 2019

Processing Big Data Across Infrastructures

Verena Kantere[⊠]

School of Electrical and Computer Engineering,
National Technical University of Athens, Athens, Greece
verena@dblab.ece.ntua.gr

Abstract. For a range of major scientific computing challenges that span fundamental and applied science, the deployment of Big Data Applications on a large-scale system, such as an internal or external cloud, a cluster or even distributed public resources ("crowd computing"), needs to be offered with guarantees of predictable performance and utilization cost. Currently, however, this is not possible, because scientific communities lack the technology, both at the level of modelling and analytics, which identifies the key characteristics of BDAs and their impact on performance. There is also little data or simulations available that address the role of the system operation and infrastructure in defining overall performance. Our vision is to fill this gap by producing a deeper understanding of how to optimize the deployment of Big Data Applications on hybrid large-scale infrastructures. Our objective is the optimal deployment of BDAs that run on systems operating on large infrastructures, in order to achieve optimal performance, while taking into account running costs. We describe a methodology to achieve this vision. The methodology starts with the modeling and profiling of applications, as well as with the exploration of alternative systems for their execution, which are hybridization's of cloud, cluster and crowd. It continues with the employment of predictions to create schemes for performance optimization with respect to cost limitations for system utilization. The schemes can accommodate execution by adapting, i.e. extend or change, the system.

Keywords: Component · Formatting · Style · Styling · Insert (key words)

1 Introduction

An ever-growing interest in analytics has made Big Data management Applications (hereafter BDAs) a number one priority that allows data-driven scientific domains to extend their research methods to an unprecedented scale and tackle new research questions, but also businesses worldwide to define new initiatives and re-evaluate their current strategies through data-driven decision-making. The deployment of BDAs on a large-scale system, such as an internal or external cloud, a cluster or the crowd, needs to be offered with guarantees of predictable performance and utilization cost. Currently, however, this is not possible, because we lack the technology that identifies the key characteristics of BDAs and their impact to performance, as well as the reasoning on the role of the system operation and infrastructure in performance. The proposed

© Springer Nature Switzerland AG 2020
S. Nepal et al. (Eds.): BIGDATA 2020, LNCS 12402, pp. 38–51, 2020.
https://doi.org/10.1007/978-3-030-59612-5_4

project aims to fill this gap by producing the research results for the development of such technology, and the technology itself.

BDAs typically produce and consume terabytes of data and they may perform complex and long-running computations. Thus, frequent access to the storage subsystem or intense use of the CPU becomes an application bottleneck. Modeling the performance with respect to I/O, CPU and memory utilization gives optimization opportunities to both system providers and users. System providers can predict such bottlenecks and utilize system combinations that may alleviate them. They can also offer cost guarantees within Service Level Agreements related to I/O or computation, as well as more appealing pricing schemes. Furthermore, the users can take informed decisions for the adaptation and the deployment of their BDA in a system, and, therefore, select in an optimal manner the computing resources needed for their application workload.

Research Objective

We need to manage efficiently the performance of BDAs that run on systems operating on large infrastructures taking into account the cost of using the underlying system. The combination of application characteristics A and system characteristics S determines the performance P of the application. Instances of the input space $I = S \times A$ are mapped to specific points or areas of P. We need to identify functions $f : I \rightarrow P$, i.e. mappings of the application and system characteristics to the application performance. We need to use these functions to predict the performance. Furthermore, the utilization of a system $s \in S$ entails a cost C, which is either the cost of power consumption or the renting price. Therefore, points or areas in P for an application $a \in A$ running on a system $s \in S$ are associated in a 1–1 manner with cost C, i.e. $(a \in, s \in S) \rightarrow (p \in P, c \in C)$. We need to manage the performance and cost by adapting either the application or the system. Based on the predictions for performance and the entailed cost, we want to create guidelines for application deployment on various system combinations with guarantees for performance, as well as cost. Ultimately, we want to select areas $I' \subseteq I$ for which performance is maximized and cost is minimized.

In this paper we describe the methodology with which we can achieve the above objective. Overall, to achieve it we have to start with the modeling of BDAs with respect to dimensions of workload, data and resources and the profiling of BDAs with respect to the proposed modeling. We also need to explore alternative systems for the deployment of BDAs, ranging from cluster, to cloud, and crowd computing, and e need to develop a methodology for the performance prediction of BDAs deployed in any of the three computing environments and combinations of them. This can be achieved with the creation of a multi-agent utilization model and the analysis of the computing environment with numerical and analytical methods. Then, we can employ the predictions to create two types of schemes for performance optimization with respect to cost limitations concerning the system utilization. The first type will reduce the execution cost by adapting, i.e. by approximating or summarizing, the workload. The second will accommodate the execution cost by adapting, i.e. extend or change, the system. Certainly, such research needs to be accompanied by thorough implementation and experimentation on real environments that can give feasible combinations of application and system characteristics. We argue that for such research we need to select at least two application environments from domains on which Big Data Analytics have and will have in the future a great impact. These two environments need to cover from end to end the profiling dimensions of underlying systems, in a very different qualitative and quantitative manner.

The efficient management of the performance of BDAs is a matter of utmost importance in the Big Data era. The research results will show the potential and the limitations of performance and cost prediction in eco-systems of users and resources from combinations cluster, cloud and crowd in arbitrary and dynamically evolving combinations. Ultimately, the results will enable the creation of policies at national and European levels concerning effective utilization of resources and constructive public engagement.

2 Related Work

The potential of managing and processing vast amounts of data in extremely large infrastructures, but, also the difficulties that such processing entails, has led both academia and industry to focus intensively on the production of such solutions. These efforts have resulted in hundreds of research papers in the last few years [1] that try to deal with the (a) heterogeneity of resources in data centers and (b) scalability problems, the (c) variability of applications and (d) unpredictability of the workload. The proposed research aspires to tackle these four issues in a holistic manner by proposing a suite of techniques, starting from the overall and generic profiling of applications on heterogeneous systems, continuing with performance and cost prediction and following up with proposals for dealing with scalability issues, based on workload approximation and summarization, as well as expansion of the underlying system and combination of it with external clouds and crowd. In the following we discuss the related work on major issues in the proposed approach, i.e. performance prediction, I/O modeling and prediction, approximation of execution and hybrid systems for large-scale computing.

Performance Prediction
Specifically, the issue of performance prediction has received great attention. The work in [2] proposes a methodology in order to predict the performance of cloud applications developed with the mOSAIC framework [3] based on benchmarking and simulation. These are much harder to achieve for any type of application that is not developed through a specific framework. Our work aims to provide a more generic methodology, that will be able to capture the characteristics of an application and predict its performance on various types of clouds, but also cluster and crowd. Furthermore, this work does not predict the scaling laws of applications. CloudProphet [4] aims at predicting the performance and costs of legacy web applications that need to be executed in the cloud, based on application instrumentation and tracing. Our work intends to predict the performance and costs for applications that can use the resource elasticity of large infrastructures. Elastisizer [5] is a system that takes as input a user query concerning the cluster size that is suitable for a MapReduce job. The system focuses on profiling in a detailed way the phases of MapReduce and their configuration parameters, e.g. how many Map and Reduce phases, how much memory is allocated to each

and what is the data access scheme, and takes into account the structure of the underlying cluster. Concerning the deployment of BDAs on large infrastructures, we need basic profiling, e.g. I/O operations and CPU usage, for general applications and not only for MapReduce: therefore, we need to focus on applications that are centralized (single-VM) or distributed (multi-VM) and process data in a batch or iterative manner.

Performance prediction has been also tackled for the specific case of database deployment in the cloud [6]. The most recent work in [7] deals with prediction through black-box modeling for similar workloads and white-box modeling for very different workloads than those observed in training. The latter gives results that are fed in statistical regression models. This work focuses on OLTP workloads and does not consider the expansion of the underlying system for accommodating the load. Rather than this, overall, such works focus on the creation of multi-tenancy strategies [e.g. 8,9] and consolidation schemes [10] for OLTP workloads and are based on the assumption that the execution time of queries is known through historical data. In our work, we will focus on big analytical workloads, for which we do not know the execution time of jobs. Such workloads differ from OLTP ones, in that they can include both short and long running tasks, which can be both I/O and CPU intensive, and can show skewness and temporal locality in data access. Therefore, we intent to profile the applications with respect to basic operations, e.g. reads and writes, as well as CPU cycles and I/O accesses, rather than profiling whole tasks, queries or transactions. Furthermore, multi-tenancy and consolidation are solutions complementary to system expansion. We intend to build policies for accommodating the application load by adding servers and/or combining outsourcing of the load to external clouds and crowd, given or adapting an existing multi-tenancy and consolidation schemes.

Performance prediction has been tackled for analytical workloads, but not extensively and not for workloads on big amounts of data. The work in [11] and its most recent continuation Contender [12], focus on batch queries that consist of a restricted number of query templates known a priori and propose a prediction technique based on simulations of the workload execution. The challenge in these works is the concurrent execution of queries. Concerning the deployment of BDAs on large infrastructures, we need we need to also tackle the challenge of workload concurrency and its effect to performance, but for a broad range of processing tasks on big amounts of data, which, furthermore, run on for large heterogeneous underlying systems.

I/O Modeling and Prediction
The sheer size of Big Data collections increases the complexity of computation, but especially of I/O management, and makes it very hard to characterize and predict the performance behavior of corresponding applications. I/O modeling and prediction have been studied in the past. In [13, 14] modeling for disk drives and disk arrays is performed. Both approaches make a hierarchical decomposition of the I/O path and examine the impact of each component separately. However, this is impossible for large-scale, complex virtualized environments. Other well-known approaches to disk array modeling are the ones presented in [15, 16] and [10]. While [15] models the disk array and [16] treats it as a black box, they both define workload dimensions and they fit samples of performance to a model. Nevertheless, as these models target only disk

arrays, the defined dimensions are not adequate to capture BDA workload characteristics. We need to fill this gap by developing methods for the thorough profiling of BDAs in terms of workload, but also, data and resources.

Although the above approaches seem to work satisfactorily enough for disk I/O prediction, they do not work when long and complex I/O subsystem paths are involved. Realizing that I/O systems become more and more complex, in [17] a self-scaling benchmark is presented. It measures I/O performance as it is seen by an end user issuing reads and writes. An end-to-end approach is also used in [18]. However, these works do not focus on applications deployed in cloud-based environments, but in parallel multicore systems; Concerning the deployment of BDAs on large infrastructures, we need to focus on the I/O and overall performance behavior of BDAs deployed in large-scale systems, with an emphasis on cloud.

I/O characterization in virtualized environments is also carried out in [19, 20]. Understanding I/O performance reveals opportunities for server consolidation and designing efficient cloud storage infrastructures. However, [19] is an experimental study for specific applications and does not include modeling of I/O behavior. An alternative approach to the problem of I/O characterization in the cloud is presented in [21]. I/O traces from production servers are collected and used as training set for a machine-learning tool. During the learning process, I/O workload types are identified automatically and as output, a I/O workload generator is produced. This generator can simulate real application workloads and thus it can be used for storage systems evaluation. Nevertheless, such an evaluation is not done yet. Concerning the deployment of BDAs on large infrastructures, we need to output results in this direction, evaluating thoroughly real cloud storage systems, but also aggregated crowd systems.

Approximate Execution

There has been a great and growing interest in the past few years on how to execute, specifically, a query workload in a way that it is approximate with respect to its actual execution, and, therefore, gain in response time.

Some of the work is on approximate query processing. The recent work in [22] as well as in [23, 24] explore querying large data by accessing only a bounded amount of it, based on formalized access constraints. These works give theoretical results on the classes of queries for which bounded evaluation is possible. Other works focus on how to pre-treat the data in order to create synopses: histograms (e.g. [25]), wavelets (e.g. [26] and sampling (e.g. [27]); or to perform execution which terminates based on cost constraints and returns intermediate results (e.g. [28]). For BDAs, it makes sense to not try to achieve approximation through alteration of the data, but through alteration of the workload.

Another type of work is on approximate query answering, in which a query that is more 'suitable' in some sense is executed in the place of the original one. In [29] a datalog program is approximated with a union of conjunctive queries, and in [22] the same example is followed with the creation of approximate versions of classes of FO queries. In a similar spirit, the works in [30, 31] deal with tractable queries for conjunctive queries and the work in [32] deals with subgraph isomorphism for graph queries. Concerning the deployment of BDAs on large infrastructures, we need to work in the same lines as these works, but focus on workflows and sets of tasks rather than queries.

Hybrid Systems for Large-Scale Computing

Issues of resource under-utilization and saturation in cloud systems can be tackled with the employment of combinations of large-scale systems that can limit over-provisioning and accommodate excessive demand in a dynamic manner. The appearance of the cloud computing paradigm was soon followed by the idea of the construction of cloud federations. The basic objective of the latter is to offer broad choices on a related set of cloud services from multiple providers. This allows cloud federations to build hybrid clouds, which can compose services from multiple sites. Currently, many cloud vendors are endorsing this idea and are building such solutions. Cisco is offering the 'Cisco Intercloud Fabric' [33], designed to combine and move, data and applications across different public or private clouds as needed. HP is launching the 'Helion Network' [34], a federated ecosystem of service providers and software vendors that will provide customers with an open market for hardware-agnostic cloud services. Amazon facilitates the migration of enterprise database legacy applications to the AWS cloud through a hybrid cloud solution that employs 'Oracle RAC on bare metal servers, either in on-premises data centers or in a private cloud, with low-latency connections to web/app tiers in AWS' [35]. Similarly, Microsoft offers the HPC Pack, which allows the creation of a hybrid high performance cluster consisting of an on-premises head node and some worker nodes deployed on-demand in Azure [36]. Even though big enterprises in cloud computing have recognized the potential of hybrid systems that consist of public clouds as well as private clouds and clusters, the research in this domain is still very recent and nascent. The work [37] is studying the dynamic allocation of resources across multiple clouds, focusing on their intertrust relationships. Similarly, [38] is proposing a way to build federations so that penalties due to possible violation of service quality by untrustworthy providers are minimized. The work in [39] presents an overall vision for the creation of migratable, self-managed virtual elastic clusters on hybrid cloud infrastructures. The work in [40] is more elaborate and proposes an online decision algorithm for outsourcing jobs, cost-based on a given budget. Beyond public and private clouds and clusters, research has not yet considered the inclusion of crowd computing in a hybrid system. The work in [41] is a first step towards this direction, and envisions volunteer cloud federations, in which clouds may join and leave without restrictions and may contribute resources without long-term commitment. The existing research results are mostly exploratory and preliminary; furthermore, they are based on cloud simulations. Concerning the deployment of BDAs on large infrastructures, we need to take this research one step further, by producing mature and complete schemes for the creation of hybrid systems that involve dynamically public and private clouds, clusters and crowd, thoroughly tested for real Big Data applications on real infrastructures.

3 Methodology for Deploying BDAs on Large Infrastructures

In order to manage the performance of BDAs we argue that we need a methodology that includes four steps: profiling of BDAs, exploration of alternative storage and processing environments, predicting performance, and, finally, optimizing the performance. In the following we give details for this methodology.

A. Profiling BDAs

The first step is to profile BDAs in order to understand their characteristics and the qualitative role they play in their performance. For this we need to create a general model for BDAs that can be employed for such characterization. In the following we will create a profiling methodology, which, based on observed measurements, will give us the characterization of a specific application with respect to the general model. We will measure and model both CPU and I/O performance.

a. Modeling BDAs

The application space A includes the following subspaces: Workload subspace, W, data subspace, D, and resource subspace, R. Thus, $A = W \times D \times R$ and points $a \in A$ represent feasible combinations of workload, data and re-source characteristics. We need to model the behavior of BDAs in terms of dimensions of W, D and R. Such dimensions may be interrelated. Our focus should be on identifying such interrelations as well as combinations of W, D and R instances that appear in realistic situations of Big Data management and have distinct impact on the performance behavior. Some of the dimensions to consider are:

Workload dimensions:

1) Size of workload in terms of number and size of processing tasks and\or number of users.
2) Type of workload in terms of: batch or iterative processing; CPU, I/O and memory intensity.
3) Data access patterns: e.g. uniform, sequential and skewed with varying degrees of skewness data access.
4) Task structure and complexity: Particular structures and complexities of tasks, for example query plans can be associated with patterns of data access and associative data access.

Data dimensions:

1) Size of the data collections accessed by the workload.
2) Replication degree and schema of the data collections.
3) The schema of the data and data dependencies.
4) The types of data: e.g. the data can be in a 'raw' format (i.e. bytestreams), or have a specific unstructured or structured formats (e.g. key-value data, RDF, relational).

Resources dimensions:
We should consider the deployment of applications that utilize either local storage or distributed object stores for the Virtual Machine (hereafter VM) storage device, as well as the deployment of applications directly on the system environment.

b. Creating a profiling methodology for BDAs

We need to create a generic process for the profiling of BDAs along the lines of the model produced in step (a). The profiling, first, samples the input space A appropriately and, second, benchmarks the BDAs in order to outline their performance behavior. The sampling takes as input the range or the set of values for the dimensions of the characteristics of the application in hand. For dimensions that contain a finite and small number of values, e.g. the type of operation {read, write}, all values can be used for sampling. For other dimensions with infinite values, e.g. the size of data, or a very large number of values, e.g. the number of VMs, some of the values will be sampled according to exploratory experiments, or pattern matching between already benchmarked generic applications. The sampling creates a subspace of A that can be used for performance measurements; acquiring these constitutes essentially the benchmarking of the BDA.

B. Exploration of alternative storage and processing environments

The second step of the methodology is to explore system architectures, i.e. instances in the system space S, that can be commonly found in environments that run BDAs and that have substantial differences, which affect the performance of BDAs. The BDAs may present very different behavior when deployed on different environments, because of the architectural complexity of the I/O path, the network connectivity and the heterogeneity of the nodes with respect to CPU and memory. We need to explore: cloud, cluster, and crowd environments.

a. Cloud computing

In a cloud environment, the nodes act as independent and need not be in the same physical location. The memory, storage device and network communication are managed by the operating system of the basic physical cloud units. This software supports the basic physical unit management and virtualization computing. We observe that the architectural complexity of the I/O path in a virtualized environment, such as the cloud, presents a serious burden in modeling the performance of a BDA. An I/O request may have to go through the VM main memory, some hypervisor-dependent drivers, the VM host memory and the network before it finally reaches the storage system, which, in turn, may have its own caches. As data centers display high heterogeneity, the structure of this path varies across different VM containers even within the same data center. Thus, a hierarchical analytical approach, as in for disk arrays, which requires a thorough evaluation and modeling of each system component separately, seems to be highly impractical: results might be too complex to analyze or combine for a definitive I/O and, therefore, BDA, modeling, especially given the numerous choices of different setups, vendors and hardware involved. In order to avoid tackling all this complexity, we need to treat the I/O subsystem path as a 'black box' and attempt to characterize it in an end-to-end fashion. Thus, we need to model I/O performance, as it is perceived by the application end-user. Since read and write operations may be fulfilled directly from caches, cache effects can be incorporated to the end-to-end I/O model.

b. Cluster computing

In a cluster environment a group of nodes are run as a single entity. The nodes are normally connected to each other using some fast local area networks. Performance and fault tolerance are the two reasons for deploying an application on a cluster rather than a single computer. In a cluster, a BDA is deployed right on the operating system, rather than in VMs. The cluster can offer parallel execution of a BDA, by employing many processors simultaneously. The cluster environment is homogeneous, as the nodes are usually in the same physical location, have the same hardware and operating system, and are directly connected with the same network. Therefore, it is possible to model the environment in more detail than the cloud. For cluster need to treat the I/O subsystem path as an 'open box', as we can have knowledge of the network connectivity and the memory of the nodes.

c. Crowd computing

A crowd environment is a combination of cloud and volunteer computing. Like cloud computing, crowd computing offers elastic, on-demand resources for the deployment of BDAs. As in cloud, the BDAs run on a virtualized environment. The nodes of the system are usually personal computers, which, naturally, are not in the same physical location. The VMs are shipped to the computers from a central location, usually a cloud or a cluster environment, and partial results produced by remote VMs are transmitted back to the central location through the Internet. The modeling of the I/O subsystem path for the crowd has the same problems as for the cloud, as the environment is once more virtualized and heterogeneous. However, in this case we have also the problem of remote access of nodes, as well as their availability fluctuations, due to their autonomy. These two factors make the crowd environment very dynamic and harder to model than the cloud. For the crowd, we need to treat the whole BDA performance in an end-to-end fashion, as perceived by the central location.

C. Performance prediction

Based on the profiling produced in step (3.1) we can develop methods for the prediction of the performance of BDAs deployed in any of the three computing environments explored in step B and combinations of them. We can achieve this with the creation of multi-agent utilization models and the analysis of the computing environment with numerical and, whenever possible, analytical methods.

a. Creation of a multi-agent utilization model

Based on the results of the methodology steps (3.1) and (3.2) we can create a multi-agent model of the utilization of the computing resources by the users. The computing systems explored in step B usually host multiple applications and/or multiple instances of the same application. The model will have the type of BDAs and number of BDA instances as well as the type and size of the system on which they are deployed as adjustable parameters. BDA instances can be described by agents whose needs are the results of a probability distribution determined by their profiling. As a simple example, we can start considering Discrete Event Simulation (DES) [42] approaches, in which the events are the requests for a running a BDA instance for a given amount of resources, i.e. a given systems. If the requested resources are available, the BDA is executed; if not it goes in a waiting queue. Upon completion, we can obtain the BDA performance and the cost of using the system.

b. Numerical and analytical analysis of the computing environment

We need to study the scaling behavior of the agent-based model under a range of conditions, to elucidate scaling laws that describe the performance and cost of sharing, amongst multiple BDAs, computing resources in cloud, cluster and crowd systems. We can vary the adjustable parameters of the multi-agent model, i.e. the type and number of BDAs, and the type and size of the system, and we can determine, through numerical simulation, mappings between the computing environment, (the application and system), and performance, i.e. functions $f : I \to P$. We can also conduct an analytical approach with simplified models, for special cases of computing environments, for example BDAs with a limited range of values for workload, data and resource dimensions, which run on a cluster.

D. Performance optimization for BDAs

The fourth step of the methodology is to employ performance predictions from step (3.3) in order create performance optimization schemes for BDAs. A monitoring process can observe the real performance of a BDA already deployed in a system and can pull performance statistics in specific time points; the monitoring schedule can be customized according to the characteristics of the specific BDA and can be decided either beforehand or on the fly, while the BDA is executing. The monitoring process can initiate and re-initiate, if needed, the profiling and performance prediction described by the methods of step (3.3). Based on measurements and predictions an advisor will suggest schemes for performance optimization. We can create two types of schemes that can improve the performance of a BDA with respect to cost limitations concerning the system utilization. The first targets to reduce the execution cost (e.g. response time or throughput) and the second to accommodate it:

a. Reduce execution cost through workload approximation and summarization

Workload Approximation: The reduction of execution cost can be done by altering the workload of the BDA. We can explore ways to approximate or summarize the workload. Workload approximation can be achieved through its relaxation from tasks that are I/O, CPU or memory intensive. Relaxed versions of the workload can be explored based on their comparison with the original version with respect to I/O, CPU or memory. We can deduce such comparisons based on the prediction functions created in step (3.3).
Workload Summarization: Beyond approximation, the execution cost can be reduced with workload summarization. We can achieve this with mining common tasks within and across workloads, and scheduling workload execution so that common tasks are

executed once or a small number of times. This is possible especially for BDAs with multiple instances, in which, frequently, workloads contain some core identical tasks and some parameterized tasks with different parameter values.

b. Accommodate execution cost through expansion and\or change of computing system

If reduction of execution cost is not a choice, we can accommodate this cost by altering the computing system. The simulation of the computing environment through multi-agent models can allow a quantitative and direct comparison of performance and utilization cost of BDAs running on cloud, both commercial and private, cluster and crowd systems. We can explore the possibility to create systems with guarantees from stochastic resources and to improve these guarantees by hybridization with dedicated resources. To achieve this we can employ the results of step (3.3), $f : I \rightarrow P$, in order to create iso quality relations between BDAs and system combinations, i.e. find functions $f_{P,C} : A' \rightarrow S'$, $A' \subset A$, $S' \subset S$ that guarantee a specific performance P with a specific cost C for a set of application instances A' and a set of system instances S'. We can use these iso relations in order to create guidelines for the online and the offline deployment of BDAs in combinations of public and private clouds, clusters and crowd systems, which guarantee the accommodation of execution cost together with some guarantees for system utilization cost. Our utmost goal should be to select application and system combinations, i.e. areas $I' \subseteq I$ for which the performance is maximized and system utilization cost is minimized.

E. Research deployment

The described research needs to be carried out in a combination of big computing environments, which, together, will allow us to investigate a very big part of the spectrum of BDA and system characteristics. In such environments we will face the challenging task of handling a big range of applications on a big range of datasets. For these, we need to collect and mine an extreme volume of time-dependent data about their operation, and produce timely profiles and predictions, which is on its own a Big Data analytics problem. As an example of the appropriateness of such environments for the deployment oft he described research, we show in Fig. 1 the complementarity of the environments of the Worldwide LHC Computing Grid (WLCG) at CERN in high-energy physics, Vital-IT (part of the Swiss Institute for Bioinformatics (SIB)) in bioinformatics, and Baobab, the high performance computing cluster of the University of Geneva. The applications of CERN and a lot of applications of Vital-IT are BDAs, in the classical sense of 4 Vs (volume, variety, velocity, veracity). The rest of the applications of Vital-IT and the applications of Baobab process large or numerous small datasets from different sources. Since all infrastructures are very big, they are heterogeneous; however, Baobab and Vital-IT have homogeneous parts.

Computing environment		CERN	Vital-IT	Baobab
Application workload	intensity	mostly I/O and CPU, and some memory	mostly I/O and some CPU	mostly memory and I/O
	processing mode	mostly batch and some interactive	interactive and batch	mostly batch
	I/O operations	variable skewness of reads/writes	variable skewness of reads and writes	range from 4 times more reads to 1000 times more writes
	data access	mostly sequential and random	mostly sequential and skewed	mostly parallel sequential
	task type	workflows of similar tasks	iterative, queries, long or short	mostly algorithms
	concurrency	medium node concurrency	medium node concurrency	high node concurrency
Application resources		single and multi-VMs	mostly physical deployment	physical deployment, deployment in VMs possible
Application data		few extremely big datasets, two replicas	numerous very big and medium datasets, some replicated	big datasets in HDF5 supporting parallel access; small text files not supporting parallel access
Storage architecture		Distributed	Local and distributed	Local and distributed

Fig. 1. Comparison of CERN, Vital-IT and Baobab

4 Conclusion

We describe our vision for research of the deployment of BDAs that run on systems operating on large infrastructures, in order to achieve optimal performance, while taking into account running costs. We propose a methodology that models and profiles applications and explores alternative systems for their execution on hybridizations of cloud, cluster and crowd. It continues with the employment of predictions to create schemes for performance optimization with respect to cost limitations for system utilization.

References

1. Jennings, B., Stadler, R.: Resource management in clouds: survey and research challenges. J. Netw. Syst. Manag. **23**(3), 567–619 (2014). https://doi.org/10.1007/s10922-014-9307-7
2. Cuomo, A., Rak, M., Villano, U.: Performance prediction of cloud applications through benchmarking and simulation. Int. J. Comput. Sci. Eng. **11**(1), 46–55 (2015)
3. Petcu, D., et al.: Architecturing a sky computing platform. In: Cezon, M., Wolfsthal, Y. (eds.) ServiceWave 2010. LNCS, vol. 6569, pp. 1–13. Springer, Heidelberg (2011). https://doi.org/10.1007/978-3-642-22760-8_1

4. Li, A., Zong, X., Kandula, S., Yang, X., Zhang, M.: CloudProphet: towards application performance prediction in cloud. SIGCOMM-Comput. Commun. Rev. **41**(4), 426 (2011)
5. Herodotou, H., Dong, F., Babu, S.: No one (cluster) size fits all: automatic cluster sizing for data-intensive analytics. In: SoCC 2011 (2011). Article no: 18
6. DBSeer: resource and performance prediction for building a next generation database cloud. In: CIDR 2013 (2013)
7. DBSeer: pain-free database administration through workload intelligence. PVLDB **8**(12), 2036–2047 (2015)
8. Zhang, Y., Wang, Z., Gao, B., Guo, C., Sun, W., Li, X.: An effective heuristic for on-line tenant placement problem in SaaS. In: ICWS, pp. 425–432 (2010)
9. Liu, Z., Hacigümüs, H., Moon, H.J., Chi, Y., Hsiung, W.-P.: PMAX: tenant placement in multitenant databases for profit maximization. In: EDBT 2013, pp. 442–453 (2013)
10. Curino, C., Jones, E.P.C., Madden, S., Balakrishnan, H.: Workload-aware database monitoring and consolidation. In: SIGMOD 2011, pp. 313–324 (2011)
11. Ahmad, M., Duan, S., Aboulnaga, A., Babu, S.: Predicting completion times of batch query workloads using interaction-aware models and simulation. In: EDBT (2011)
12. Duggan, J., Papaemmanouil, O., Çetintemel, U., Upfal, E.: Contender: a resource modeling approach for concurrent query performance prediction. In: EDBT 2014, pp. 109–120 (2014)
13. Ruemmler, C., Wilkes, J.: An introduction to disk drive modeling. IEEE Comput. **27**(3), 17–28 (1994)
14. Uysal, M., Alvarez, G.A., Merchant, A.: A modular analytical throughput model for modern disk arrays. In: MASCOTS (2001)
15. Anderson, E.: Simple table-based modeling of storage devices. Technical report, HP Labs (2001)
16. Wang, M., Au, K., Ailamaki, A., Brockwell, A., Faloutsos, C., Ganger, G.R.: Storage device performance prediction with CART models. In: MASCOTS (2004)
17. Chen, P., Patterson, D.A.: A new approach to I/O performance evaluation-self scaling I/O benchmarks, predicted I/O performance. In: SIGMETRICS (1993)
18. Ipek, E., de Supinski, B.R., Schulz, M., McKee, S.A.: An approach to performance prediction for parallel applications. In: Cunha, J.C., Medeiros, P.D. (eds.) Euro-Par 2005. LNCS, vol. 3648, pp. 196–205. Springer, Heidelberg (2005). https://doi.org/10.1007/11549468_24
19. Gulati, A., Kumar, C., Ahmad, I.: Storage workload characterization and consolidation in virtualized environments. In: VPACT (2009)
20. Kraft, S., Casale, G., Krishnamurthy, D., Greer, D., Kilpatrick, P.: Performance models of storage contention in cloud environments. Softw. Syst. Model. **12**(4), 681–704 (2013). https://doi.org/10.1007/s10270-012-0227-2
21. Delimitrou, C., Sankar, S., Vaid, K., Kozyrakis, C.: Decoupling datacenter studies from access to large-scale applications: a modeling approach for storage workloads. In: IISWC (2011)
22. Potti, N., Patel, J.M.: DAQ: a new paradigm for approximate query processing. PVLDB **8**(9), 898–909 (2015)
23. Fan, W., Geerts, F., Libkin, L.: On scale independence for querying big data. In: PODS (2014)
24. Cao, Y., Fan, W., Yu, W.: Bounded conjunctive queries. PVLDB **7**(12), 1231–1242 (2014)
25. Jagadish, H.V., Koudas, N., Muthukrishnan, S., Poosala, V., Sevcik, K.C., Suel, T.: Optimal histograms with quality guarantees. In: VLDB (2009)
26. Garofalakis, M.N., Gibbons, P.B.: Wavelet synopses with error guarantees. In: SIGMOD (2004)

27. Agarwal, S., et al.: Knowing when you're wrong: building fast and reliable approximate query processing systems. In: SIGMOD (2014)
28. Agarwal, S., Mozafari, B., Panda, A., Milner, H., Madden, S., Stoica, I.: BlinkDB: queries with bounded errors and bounded response times on very large data. In: EuroSys (2013)
29. Chaudhuri, S., Kolaitis, P.G.: Can datalog be approximated? JCSS 55(2), 355–369 (1997)
30. Barcelo, P., Libkin, L., Romero, M.: Efficient approximations of conjunctive queries. SICOMP 43(3), 1085–1130 (2014)
31. Fink, R., Olteanu, D.: On the optimal approximation of queries using tractable propositional languages. In: ICDT (2011)
32. Fan, W., Li, J., Ma, S., Tang, N., Wu, Y., Wu, Y.: Graph pattern matching: from intractability to polynomial time. PVLDB 3(1), 1161–1172 (2010)
33. http://www.cisco.com/c/en/us/products/cloud-systems-management/intercloud-fabric/index.html
34. http://www8.hp.com/us/en/cloud/helion-network-overview.html
35. https://reinvent.awsevents.com/files/sponsors/Logicworks_Hybrid_Cloud_Legacy_Applications_WP.pdf
36. https://technet.microsoft.com/en-us/library/jj899572.aspx
37. Lo, N.-W., Liu, P.-Y.: An efficient resource allocation framework for cloud federations. J. Inf. Technol. Control 44(1) (2015)
38. Hassan, M.M., Alelaiwi, A., Alamri, A.: A dynamic and efficient coalition formation game in cloud federation for multimedia applications. In: GCA (2015)
39. Calatrava, A., Moltó, G., Romero, E., Caballer, M., de Alfonso, C.: Towards migratable elastic virtual clusters on hybrid clouds. In: IEEE CLOUD (2015)
40. Niu, Y., Luo, B., Liu, F., Liu, J., Li, B.: When hybrid cloud meets flash crowd: towards cost-effective service provisioning. In: IEEE INFOCOM (2015)
41. Rezgui, A., Rezgui, S.: A stochastic approach for virtual machine placement in volunteer cloud federations. In: IEEE IC2E (2014)
42. Pllana, S., Fahringer, T.: Performance prophet: a performance modeling and prediction tool for parallel and distributed programs. In: ICPP Workshops (2005)

Fake News Classification of Social Media Through Sentiment Analysis

Lixuan Ding, Lanting Ding, and Richard O. Sinnott$^{(\boxtimes)}$

School of Computing and Information Systems, University of Melbourne,
Melbourne, Australia
rsinnott@unimelb.edu.au

Abstract. The impacts of the Internet and the ability for information to flow in real-time to all corners of the globe has brought many benefits to society. However, this capability has downsides. Information can be inexact, misleading or indeed downright and deliberately false. *Fake news* has now entered the common vernacular. In this work, we consider fake news with specific regard to social media. We hypothesise that fake news typically deals with emotive topics that are deliberated targeted to cause a reaction and encourages the spread of information. As such, we explore sentiment analysis of real and fake news as reported in social networks (Twitter). Specifically, we develop an AWS-based Cloud platform utilising news contained in the untrustworthy resource *Fake-NewsNet* and a more trusted resource *CredBank*. We train algorithms using Naive Bayes, Decision Tree and Bi-LSTM for sentiment classification and feature selection. We show how social media sentiment can be used to improve the accuracy in identification of fake news from real news.

Keywords: Fake news · Sentiment analysis · Fact checking

1 Introduction

Social media has become one of the most commonplace sources of information for humans. Social media is used increasingly as a prime source of news and current affairs. However, social media is increasingly considered as a double-edged sword with regards to the way it can be used for the near real-time, global spreading of information. Whilst it allows users to share and consume information in an easy and user-tailored manner, there are typically no checks on what information is sent. Indeed, it is possible to produce fake and harmful news resulting in spreading of low quality or incorrect information [1]. This rapid spreading of fake news can cause enormous harm to individuals, society and even to countries. In this context, public sentiment has an increasing influence on society. Understanding public opinion is commonly demanded for many areas: branding of products, through to predicting events such as voting outcomes. The ability to accurately gauge public sentiment in near real-time is thus highly desirable. Sentiment analysis is one approach for establishing public sentiment. In this work, we consider the role of sentiment analysis and the way it may be used to distinguish real and fake news to help unlock some of the challenges related to unchecked fake news distribution.

© Springer Nature Switzerland AG 2020
S. Nepal et al. (Eds.): BIGDATA 2020, LNCS 12402, pp. 52–67, 2020.
https://doi.org/10.1007/978-3-030-59612-5_5

As one of the most popular mainstream social networks that has attracted a huge number of users, Twitter has established itself as an important platform for the distribution of news: be the content fake or real. In this project, we utilise two Twitter news data sets: *FakeNewsNet* [5] and *CredBank* [18] reflecting real and fake news, and official (true) resources respectively. *FakeNewsNet* includes news categories related to politics and gossip, each of which provides Tweet IDs for real and fake news. *CredBank* comprises many topics and tweets that have been (manually) classified as real news. The total amount of data used was over 2.5 million tweets.

To tackle the classification of sentiment, this paper uses traditional Natural Language Processing (NLP) technology to process these corpora and applies two different machine learning methods to perform sentiment analysis and automatic discrimination/prediction of real or fake news as reported in social media networks, namely: Naive Bayes and Decision Trees [19]. The sentiment analysis undertaken focuses on the sentiment tendency of Tweet texts, the word polarity in texts, and the sentiment characteristics of language features. To tackle big data problems, we leverage machine learning approaches including Bi-Directional Long-Short Term Memory (Bi-LSTM) [13] to train real and fake news classifiers to test whether real and fake news could be distinguished based on their sentiment and importantly on their associated content.

2 Related Work

As an important medium for information dissemination, the information quality in social media has been explored by many researchers with extensive studies based on social network data and data analysis. Capturing the sentiment of posts has also been extensively explored [20]. This has included non-English sentiment analysis [2] as well as combinations of geographical location, time factor and sentiment analysis, e.g. changing sentiment in different locations at different times in New York City [3].

In this work we focus on real and fake news of short messages (tweets). Buntain and Golbeck (2017) [4] developed a Twitter data set (*CredBank*) targeted on the accurate assessment of news as shown on Twitter. Shu et al. (2019) provided a fake and real news data set (*FakeNewsNet*) collected from Twitter including two comprehensive data sets related to politics and gossip. In this work, each post contained news content, data sources and other information such as images and videos.

For sentiment analysis, many works focus on text pre-processing, feature extraction, and model training. For example, [6] use traditional natural language processing methods to establish a framework for sentiment classification with two key elements: a classifier and a feature extractor. In their work, classifiers were trained using machine learning models including Naive Bayes, Maximum Entropy, and Support Vector Machines (SVM). The feature extractor utilized N-Gram methods including unigram and bigrams. Naradhipa et al. [2] used the Levensthein distance to normalize the word format and divide the sentiment analysis into two types: *sentence-level* and *word-level*. As with many approaches, the sentiments were classified into positive, negative, and neutral. They used SVM with Maximum Entropy as their classification algorithm.

Buntain and Golbeck extracted 45 features from social media and divided these features into four categories: *structural features, user features, content features* and *temporal features* [4]. Shu et al. (2017) also extracted features related to different aspects of tweets, including news content-based features, language vocabulary-based features, visual-based features, social context-based features, and time sequence-based features. Tang et al. (2014) [7] proposed a new method for learning word-embeddings suited for sentiment analysis termed sentiment-specific word embedding. By consideration of the position of the content, this method addressed many of the shortcomings of traditional continuous word vector establishment methods, where only the establishment of vectors with similar sentiment word polarity was considered. They implemented this method and used it for training different neural networks.

Whilst various researchers have explored fake news and sentiment analysis, thus far no work has attempted to combine these two approaches. This is the contribution of this paper. Importantly, the work requires the development of a large-scale data processing platform capable of dealing with classification, feature extraction and sentiment analysis at scale.

3   System Architecture and Hypotheses

Social media data analytics, and especially the application of machine learning approaches demand flexible and scalable infrastructure. In this work, the Amazon Web Service (AWS) Cloud platform was chosen as the infrastructure provider as it has a diverse range of stable services and features. Figure 1 illustrates the high-level architecture that was utilized in this work including the Twitter social media harvesters, the storage systems and the data processing and analysis components. In this, figure, the grey lines indicate the data flow through the system.

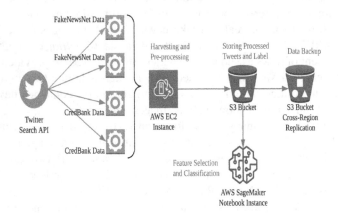

Fig. 1. System architecture.

In this work, the harvesters ran on the AWS Elastic Compute Cloud (EC2) platform with four processes running in parallel to harvest tweets through the Twitter Search API. Two processes were dedicated to harvesting tweets from *CredBank* and *FakeNewsNet* by matching the Tweet IDs provided in the dataset. The harvested tweets were stored locally on each instance where they could be retrieved for pre-processing. The pre-processed tweets together with the labels specified in the two datasets were then used by the analyser.

Since the project only focused on Twitter (tweet) texts and their labels, the data that was required to be stored had a relatively simple structure. Therefore, the AWS Simple Storage System (S3) was selected as the back-end storage system. S3 buckets were established with one as the primary storage, the other provided cross-region replication and acted as the backup. The pre-processed tweets were stored as five large JSON files in buckets based on their associated categories. The analyser was implemented using AWS SageMaker, a dedicated machine learning platform that provides an integrated Jupyter notebook instance [8]. The analyser retrieved data from S3 buckets using the Python Boto library. It then performed the sentiment analysis and classification tasks, before storing the results.

Allcott and Gentzkow identified that the diffusion pattern of false news goes significantly further, faster, deeper, and broader than more truthful (non-fake) news [9]. The reasons behind this can be multi-fold, e.g. an artefact of human nature or based on a penchant for conspiracy. This applies to many domains and contexts. In the finance domain, for example, finance-based fake news is often offered as "clickbait" for advertisements that are profit-driven and tasked with attracting website traffic [10]. This can trigger a stronger emotional response that lends itself as a trigger for further spreading of disinformation. Conspiracy-based fake news focuses on promoting particular ideas, advancing certain agendas, or discrediting others. Again, such information is intended to elicit strong emotional responses. Political and fake gossip news is a common type of conspiracy-based fake news. As a mainstream news and social networking service, Twitter is recognized as a major spreader of fake news. While Tweet texts are generally not the sole source of fake news, e.g. they may include links to websites containing the erroneous information. Through these links, users may be misled to spreading such tweets and hence disseminating the fake news. As such, the strength of emotion that fake news can trigger is important. Fake news that no-one cares about is unlikely to be redistributed (retweeted). Based on this, we focus on three key considerations:

- Establishing the sentiment of fake and real news as recorded in tweets;
- Understanding the characteristics and textual features of tweets that can potentially be used to distinguish between real and fake news, and finally
- Whether we can actually distinguish fake and real news tweets based on their content.

In exploring these considerations, we consider two underlying hypotheses regarding the sentiment of real and fake news:

- **Hypothesis 1.** *Fake news tweets are expected to include more emotive content, i.e. more positive or more negative content;*

- **Hypothesis 2.** *Language features can be used to classify and hence distinguish real and fake news.*

4 Methodology

To explore these hypotheses, we apply traditional machine learning methods and a neural network trained as a classifier incorporating the sentiment analysis of fake and real news in social media. *FakeNewsNet* contains data sets from both BuzzFeed and the PolitiFact [21] platforms, including two categories of news: *politics* and *gossip*. These data sets were provided as four different CSV files: *politifact_fake.csv*; *politifact_real. csv*; *gossipcop_fake.csv* and *gossipcop_real.csv*. Each CSV file had four columns: -id: identifying each news uniquely; -URL: the URL link of original news posted on the network; -title: the news article title, and the -tweet_ids: the unique Tweet ID related with the tweet used to share the news.

The *CredBank* data also includes a Credibility Annotation File with 1,377 topics where each topic was rated by 30 independent and external public assessors with scores ranging from −2 to +2, reflecting the credibility of the news from low to high together with the corresponding reasons for the score [4]. Each topic comprised numerous tweet ids. The average ratings of the assessors and the overall credibility of the tweets was used for the individual topics. In this work we selected 88 topics with average ratings above zero, i.e. the tweets could be considered as real news. The total number of real and fake news used is shown in Table 1.

Table 1. Data distribution

News type	Fake news	Real news	Total
FakeNewsNet Politics	150,064	421,034	571,098
FakeNewsNet Gossip	529,435	837,554	1,366,989
CredBank	0	499,277	499,277
Total	679,499	1,757,865	2,437,364

Having harvested social media data, it was necessary to process the original data since there could be redundant information. Removing replicas is a typical first step in data processing. Since news collected from *FakeNewsNet* is not always English news, removing non-English content is important. The Python library *langid* was used to determine the tweet language. All real news was initially labelled "0", while fake news was labelled as "1". The processed data was stored in compressed JSON format. Each data comprised two key-value pairs including "text" for news content and "label" for the news class. After pre-processing, the valid news articles were reduced with results shown in Table 2.

Table 2. Data distribution after pre-processing.

	Fake news	Real news	Total
FakeNewsNet Politics	62,512	258,771	321,283
FakeNewsNet Gossip	291,635	298,494	590,129
CredBank	0	499,277	499,277
Total	354,147	1,056,542	1,410,689

In order to more efficiently conduct sentiment analysis on fake and real news in social media, it is essential to remove unnecessary information, such as URL links, stop words, emojis and punctuation. Regular expressions were used for data cleaning and the Python library *spaCy* used for word tokenization and lemmatization. This library also facilitates the identification of relevant features.

However, as can be seen from Table 3, the distribution of news data is not balanced which can cause errors when training the models. Therefore, sampling methods were utilised including under-sampling and over-sampling. Under-sampling is used to randomly sample news data for each kind of news, while *FakeNewsNet* comprises significantly less data. Thus, over-sampling was used to balance the disparity between the amount of fake and real news. The Python library *imblearn* provides an algorithm (SMOTE) for solving such unbalanced data issues. To test the method, we collected news samples in initial experiments and identified that the classifiers performed better when trained by over-sampling the data compared to under-sampling. As a result, over-sampling was adopted as the method.

Table 3. Performance of sampling methods.

Sampling method	Naive Bayes	decision tree
Non-Ov-sampling (60000:60000)	0.77	0.85
Over-sampling (60000:10000) – > (60000:60000)	0.75	0.81
Under-sampling (10000:10000)	0.65	0.71

Feature engineering and especially feature selection play a very important role in machine learning. Feature selection is used to deal with data dimension issues with sparse data samples or where there is difficulty in calculating Levenshtein distances. Feature selection removes unrelated features which can help to speed up model training, reduce training vectors and improve the overall model performance.

There are many algorithms for feature selection, such as removing features with low variance, univariate feature selection, and loop feature selection [22]. In this work, the F-value belonging to univariate feature selection methods was used as part of the F-test to determine the features with high correlation rates between features and their dependent variables. The Scikit-learn in Python provides a method *"f-classif"* supporting this algorithm and was applied.

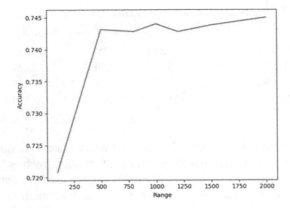

Fig. 2. Model performance with changing of numbers of features.

The feature number is also a critical factor affecting the performance of machine learning models. In this work, 10-fold cross-validation was used to evaluate the performance of the model with different numbers of features. Figure 2 illustrates one model (Decision Tree) trained with 2000 features. This model performed best in this project.

Naive Bayes was trained as the baseline, and the pre-processed data set then utilised for training the decision tree classifier. Both of the classifiers mentioned above were trained with methods that Python Scikit-learn provides. Having considered that different models perform differently on different news data sets, the training and test data sets were divided into three types: datasets with real and fake news for politics only, datasets with real and fake news for gossip only, and datasets with real and fake news comprising mixed news categories. During the training process, classifiers were used to classify the data into two categories: fake news and real news. After the model training was complete, the test data sets were evaluated for *Accuracy*, *Precision* and *Recall* as evaluation metrics to detect the performance of the trained models. Table 4 shows the results of the different approaches.

Table 4. Performance of different methods trained with different training datasets.

News type	Method	Accuracy	Precision	Recall
Gossip	Naïve Bayes	0.72	0.79	0.72
	Decision tree	0.74	0.76	0.74
Politics	Naïve Bayes	0.71	0.75	0.71
	Decision tree	0.76	0.81	0.76
Combined	Naïve Bayes	0.67	0.75	0.67
	Decision tree	0.68	0.73	0.68

As can be seen from Table 4, the decision tree approach generally performed better than Naive Bayes. A part of the decision tree showing the construction of the decision tree models is shown in Fig. 3(a) where each node reflects one feature acting on the decision tree. The thumbnails of the decision tree classifiers as trained by the different training sets are also shown in Fig. 3(b).

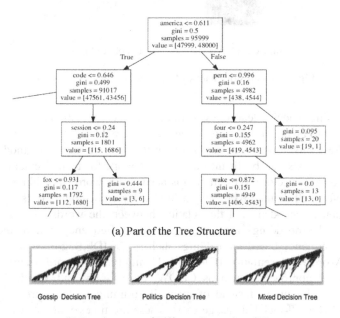

(a) Part of the Tree Structure

Gossip Decision Tree   Politics Decision Tree   Mixed Decision Tree

(b) Tree Structure of Different Categories of News

Fig. 3. Decision tree construction

As can be seen, there are fewer branches in the politics decision tree, i.e. the decision tree can more readily classify real news and fake news for political news. In contrast, the decision tree trained by mixing data sets performed poorly with many more branches.

In order to compare the prediction results between different categories of news, a confusion matrix was used as the evaluation method. This reflects the true condition of classifiers directly. Figure 4 shows the results of the confusion matrix. For the predictive classification model, the more accurate the prediction, the better the classifier. Here the second and fourth quadrants visualise the number of True Positive and True Negative objects respectively. As seen, it is more probable to predict fake news correctly for gossip news, whilst it is more probable to predict real news correctly for political news.

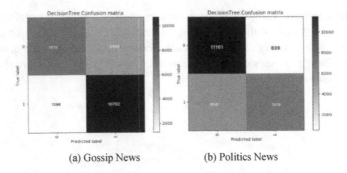

(a) Gossip News   (b) Politics News

Fig. 4. Confusion matrix of different categories of news.

As seen, using machine learning generates a large computational matrix, which can make it difficult to be utilized at larger scale, i.e. with real fake news scenarios using big data. To address this, a neural network was used as an alternative method.

Before input texts can be fit into a neural network for textual entailment, a conversion of text to its vector representation is needed, i.e. word embedding, since neural networks prefer well defined fixed-length inputs. By embedding words in documents, the information is encoded, and the relation between the words can be established. Typically, word embeddings are categorised using frequency and prediction-based approaches. Frequency-based embeddings such as TF-IDF,Bag-of-Words and Global Vectors for Word Representation (GloVe) [11] simply count word occurrences, while prediction-based approaches such as CBOW, Skip Gram, and Word2Vec take the context into consideration. Instead of manually training vector representations, pre-trained models for off-the-shelf usage exist. These are timesaving, statistically robust, and can leverage massive corpora. Among those models, two of the most popular ones are *Word2Vec* and *GloVe*. In Glove, word-word co-occurrence probability and unsupervised learning are used [11]. In this work, GloVe was chosen for word embedding as the initial weight for the neural network classification. It captures sub-linear relationships between two words in the same position in the vector space. It has been shown to have better performance in word analogy tasks. For deeper analysis, machine learning approaches were also applied.

LSTM is a variation of recurrent neural networks (RNN) [13]. An RNN takes both current and the recent past steps as inputs and combines them to determine how to respond to new data through a feedback loop. In this way, it behaves like memory. To find the correlation between input sequences through multiple steps or long-term dependencies, each input is stored in RNN's hidden state and is cascaded forward to affect processing of the inputs in each step [12]. The aim of training an RNN model is to minimize the difference between the output and the correct label using given inputs by adjusting a weight matrix through an optimizer and loss functions.

Unlike traditional RNNs, where long-term contexts can be neglected due to the vanishing gradient problem on back propagation, LSTMs are able to preserve the error and learn through remote contexts using multiple switch gates. Bidirectional LSTMs (Bi-LSTMs) train two instead of one LSTM on an input sequence, with the first one being the original sentence embeddings, and the second being the reversed copy. This can provide context to the left and right and result in faster and richer learning on language problems [13]. Bi-LSTMs also have a relatively good performance for classifying sequential patterns as often arise in texts [14].

In this work, the Python Keras library with TensorFlow as a backend was chosen as the implementation tool with the input, hidden, and output layers configured as follows. In the *input layer*, each pre-processed tweet text was set as a single input. Tokenization was performed using Keras Tokenizers for sequence embeddings with length padded to 100 to provide uniformity. The padded embeddings were then fed into each neuron as data input. The initial weights of the model were set using the pre-trained GloVe embeddings of the input sentences, where words were converted into vectors according to their frequencies in documents and position in sentences. The *hidden layer* consisted of a Bi-LSTM Layer and two Dense Layers with the parameters as listed in Table 5.

Table 5. Hidden layer parameters setting

Activation function	Loss Function	Optimizer	Dropout
Rectified linear unit (ReLU)	Binary Cross-Entropy	Adam	0.5

The Activation Function *Rectified Linear Unit* (ReLU) was selected as the activation function to transform the weighted sums to activation in the hidden layer to overcome vanishing gradient problems. A *Binary Cross-Entropy* function was chosen since the output class is binary. This saves memory as it does not require the target variable to be one-hot encoded prior to training. The *Adam* optimizer was used to update the weights and minimizing the loss function since it has faster convergence [16]. The *dropout* used to prevent overfitting and to improve generalization error by randomly ignoring some outputs was set to 0.5 for the Bi-LSTM layer. This is close to the optimal experimental value for a wide range of networks and tasks as shown by [17]. In the *output layer*, a soft-max activation function was used in an Output Dense Layer to aggregate the results and to give the final probabilities.

As is described previously, tweets were collected related to different topics. In order to analyse whether fake news could be detected purely based on the content, the influence of the topic must be considered. To tackle this, the datasets were partitioned in two different ways. In the first partition, tweets from *FakeNewsNet* and *CredBank* were mixed and then divided to form training and testing sets to test the performance of the news topics in the training set. The second partition takes only the *FakeNewsNet* data as the training set and *CredBank* Real News as the test set to test the performance when related news topics are new to the classifier. The number of partitioned tweets is shown in Table 6. It is noted that the number of fake news collected in *CredBank* is zero since the average ratings are chosen to judge whether the news is real or fake, and only a few of the news tweets had an average rating below zero. If those news items were added, the dataset would be seriously unbalanced.

Table 6. Training and testing dataset partition settings.

Label				
Type	Real news		Fake news	Total
	Fake News Net	CredBank	Fake NewsNet	
(a) Mixed partition settings on training and testing Set				
Training set	150,000	359,277	160,000	659277
	509,277			
Testing set	160,000	150,000	150,000	460000
	310,000			
(b) Separated partition settings on training and testing datasets				
Training set (FakeNewsNet)	310,000		310,000	620,000
Testing set (CredBank)	499,277		NA	499,277

Table 7. Sentiment portion in real and fake news.

Attributes	Gossip	Political	Combined
Real news	P (%) > N (%)	P (%) > N (%)	P (%) > N (%)
Fake news	P (%) > N (%)	N (%) > P (%)	P (%) > N (%)
Positive sentiment (%)	R > F	R > F	R > F
Negative sentiment (%)	F > R	F > R	F > R

(P: POSITIVE, N: NEGATIVE, R: REAL NEWS, F: FAKE NEWS)

5 Results and Analysis

In order to analyze the sentimental polarity of the tweet text and the words, the Python libraries *textblob* and *nltk.corpus* were used.

To explore Hypothesis 1 (*Fake news tweets are expected to include more positive or more negative content*), the sentiment analysis on the *FakeNewsNet* data set involved classifying the tweets based on their sentiment, i.e. whether they expressed a positive, negative or neutral sentiment. The results are shown in Fig. 5.

From Fig. 5 above, it can be seen that for all types of news tweets, emotional (positive and negative) content take up a larger portion compared to neutral content, which suggests Hypothesis 1 is not true. However, if we investigate the relationship between positive and negative sentiments in more detail, it can be observed that the different types of news exhibit different patterns. The relationship between the portion of positive and negative sentiments in real and fake news tweet, and the portion of these two sentiments types in real news tweets, when compared to fake news tweets, is shown in Table 7. As seen in Table 7 for all of the three types of real news tweets, positive content takes up a larger portion compared to negative content, while for fake news tweets, only political news related tweets have more negative sentiment than

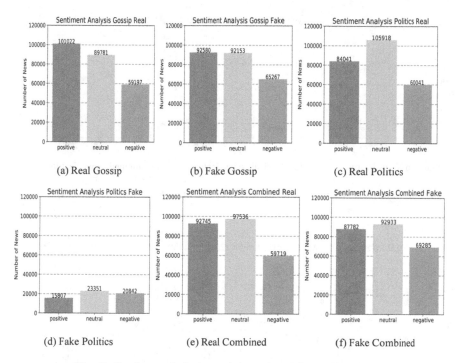

Fig. 5. Sentiment analysis on tweet content (FakeNewNet Dataset).

positive. This suggests that sentiment analysis could help to detect political fake news as it presents a different pattern compared to real news. For all these three types of news, positive sentiment takes up a larger portion of real news compared to fake news, while the percentage of negative sentiment is larger in fake news. Thus, it can be considered that more real news tweets tend to have positive sentiment, while more fake news Tweets present negative sentiment. Note that the result may be biased based on the number and balance of real and fake news tweets in the three categories.

To explore Hypothesis 2 (*Language features can be used to classify and hence distinguish real and fake news*) the sentiment distribution of the top 2000 features that best distinguish fake and real news tweets for gossip, political news and combinations of both are illustrated in Fig. 6.

It can be observed from Fig. 6 that the majority of features are neutral. Positive and negative features only take up a small proportion of the features, and their number is (relatively) similar across all the three categories, which suggests Hypothesis 2 is not true. To investigate the characteristics of good features, 10 features with the highest contribution to the decision tree for real and fake news Tweets of each category were considered and plotted as shown below in Fig. 7.

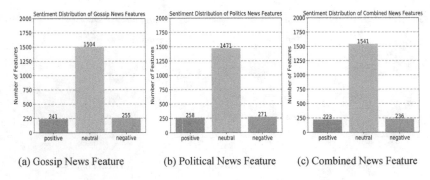

(a) Gossip News Feature   (b) Political News Feature   (c) Combined News Feature

Fig. 6. Sentiment distribution of the top 2000 features (FakeNewsNet dataset).

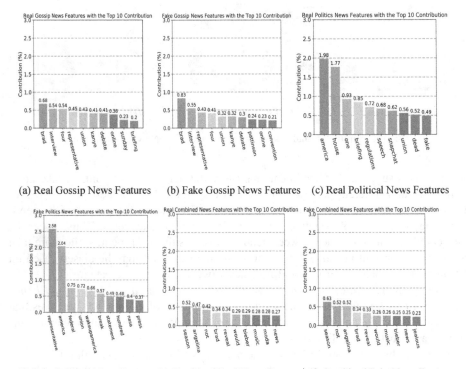

(a) Real Gossip News Features   (b) Fake Gossip News Features   (c) Real Political News Features

(d) Fake Political News Features (e) Combined Real News Features   (f) Combined Fake News Features

Fig. 7. Top 10 features with best contributions (FakeNewsNet dataset).

As seen from Fig. 7, features with the largest contributions are related to news topics selected in the *FakeNewsNet* dataset. The most typical feature in gossip news is names such as 'Brad' and 'Angelina'. For political news, words typically refer to organizations such as 'NASA' or 'press'. Although there are correlations between these features, the predictability of these features cannot be guaranteed, and new topics may be included in other scenarios and in other contexts. Another observation is that the contribution of political fake news features is relatively higher than other categories, which may imply that political fake news can more easily be detected. The Bi-LSTM classification results in two differently partitioned data sets as shown in Table 8:

Table 8. Bi-LSTM classification results.

News Type	Separated	Mixed
Real news tweets	36.6%	97.2%
Fake news tweets	NA	93.8%

As can be seen from Table 8, mixing training and testing sets has a much better performance than separated ones. This indicates that the prediction ability of Twitter content is highly dependent on the topics identified in the training set. In the real world, when an unrelated (new) topic is introduced, the classifier would be more likely to give a wrong prediction if only the news content was used.

It is noted that the limitation of this test is that it only includes the real news in a separated test set, as only a few topics were classified to be fake on average in the *CredBank* data set. However, as the difference between separated and mixed partition on real news classification results is considerable, it can be deduced that the situation for fake news would not be much better.

6 Conclusions and Future Work

This work focused on the analysis of fake and real news through social media using decision trees as the classifier to train and establish the contributions of features. A Bi-LSTM model was used as a classifier with two differently partitioned training and testing sets used to analyze whether fake news could be detected purely based on its content. It was discovered that for all types of news related tweets, emotional content takes up a larger portion compared to neutral content, with more real news tweets tending to have a more positive sentiment, and more fake news tweets having a more negative sentiment. Of the three categories, political news was the only one that presented different sentiment patterns for fake and real news-related tweets. Although sentiment analysis could help to detect fake political news, the majority of the features that distinguish them was shown to be neutral. Typical patterns included in features with the best contributions included individual names and words related to organizations. However, as these features are all topic-related, it was identified that the prediction ability of Twitter content was highly dependent on the topics it had seen before, which was reinforced through the results of Bi-LSTM classifications.

The source code for this work is available at https://github.com/lixuand1/Fake-news-and-social-media/tree/master/code.

References

1. Shu, K., Sliva, A., Wang, S., Tang, J., Liu, H.: Fake news detection on social media. ACM SIGKDD Explor. Newslett. **19**(1), 22–36 (2017). https://doi.org/10.1145/3137597. 3137600
2. Naradhipa, A.R., Purwarianti, A.: Sentiment classification for Indonesian message in social media. In: 2012 International Conference on Cloud Computing and Social Networking (ICCCSN) (2012). https://doi.org/10.1109/icccsn.2012.6215730
3. Bertrand, K., Bialik, M., Virdee, K., Gros, A., Bar-Yam, Y.: Sentiment in New York City: A High Resolution Spatial and Temporal View (2019). https://arxiv.org/abs/1308.5010. Accessed 21 Nov 2019
4. Buntain, C., Golbeck, J.: Automatically identifying fake news in popular twitter threads. In: 2017 IEEE International Conference on Smart Cloud (Smartcloud) (2017). https://doi.org/10.1109/smartcloud.2017.40
5. Shu, K., Mahudeswaran, D., Wang, S., Lee, D., Liu, H.: FakeNewsNet: a data repository with news content, social context and spatialtemporal information for studying fake news on social media (2019). https://arxiv.org/abs/1809.01286. Accessed 21 Nov 2019
6. Go, A., Bhayani, R., Huang, L.: Twitter Sentiment Classification using Distant Supervision [Ebook], 1st edn. Stanford (2009). https://cs.stanford.edu/people/alecmgo/papers/TwitterDistantSupervision09.pdf
7. Tang, D., Wei, F., Yang, N., Zhou, M., Liu, T., Qin, B.: Learning sentiment-specific word embedding for twitter sentiment classification. In: Proceedings of the 52nd Annual Meeting of the Association for Computational Linguistics (Volume 1: Long Papers) (2014). https://doi.org/10.3115/v1/p14-1146
8. What Is Amazon SageMaker? - Amazon SageMaker. (2019). https://docs.aws.amazon.com/sagemaker/latest/dg/whatis.html. Accessed 21 Nov 2019
9. Allcott, H., Gentzkow, M.: Social media and fake news in the 2016 election. J. Econ. Perspect. **31**(2), 211–236 (2017). https://doi.org/10.1257/jep.31.2.211
10. Mills, A., Pitt, C., Ferguson, S.: The relationship between fake news and advertising. J. Advertising Res. **59**(1), 3–8 (2019). https://doi.org/10.2501/jar-2019-007
11. Pennington, J.: GloVe: Global Vectors for Word Representation (2019). https://nlp.stanford.edu/projects/glove/. Accessed 21 Nov 2019
12. Goodfellow, I., Bengio, Y., Courville, A.: Deep Learning, 1st edn. pp. 373–375. Cambridge (2016)
13. Mohs, R., Wescourt, K., Atkinson, R.: Effects of short-term memory contents on short-and long-term memory searches. Memory Cognit. **1**(4), 443–448 (1973). https://doi.org/10.3758/bf03208906
14. Makarenkov, V., Rokach, L., Shapira, B.: Choosing the right word: using bidirectional LSTM tagger for writing support systems. Eng. Appl. Artif. Intell. **84**, 1–10 (2019). https://doi.org/10.1016/j.engappai.2019.05.003
15. Yildirim, Ö.: A novel wavelet sequence based on deep bidirectional LSTM network model for ECG signal classification. Comput. Biol. Med. **96**, 189–202 (2018). https://doi.org/10.1016/j.compbiomed.2018.03.016
16. Kingma, D., Ba, J.: Adam: A Method for Stochastic Optimization (2019). https://arxiv.org/abs/1412.6980v8. Accessed 21 Nov 2019
17. Poernomo, A., Kang, D.: Biased dropout and crossmap dropout: learning towards effective dropout regularization in convolutional neural network. Neural Netw. **104**, 60–67 (2018). https://doi.org/10.1016/j.neunet.2018.03.016

18. Mitra, T., Gilbert, E.: Credbank: a large-scale social media corpus with associated credibility annotations. In: Proceedings of the Ninth International AAAI Conference on Web and Social Media (2015)
19. Caruana, R., Niculescu-Mizil, A.: An empirical comparison of supervised learning algorithms. In: Proceedings of the 23rd International Conference on Machine Learning - ICML 2006 (2006). https://doi.org/10.1145/1143844.1143865
20. Li, Z., Fan, Y., Jiang, B., Lei, T., Liu, W.: A survey on sentiment analysis and opinion mining for social multimedia. Multimed. Tools Appl. **78**(6), 6939–6967 (2018). https://doi.org/10.1007/s11042-018-6445-z
21. Wang, W.: Liar, Liar Pants on Fire: A New Benchmark Dataset for Fake News Detection (2019). https://arxiv.org/abs/1705.00648. Accessed 21 Nov 2019
22. Sulaiman, M., Labadin, J.: Feature selection based on mutual information. In: 2015 9th International Conference on IT in Asia (CITA) (2015). https://doi.org/10.1109/cita.2015.7349827
23. Sha, L., Chang, B., Sui, Z., Li, S.: Reading and thinking: re-read LSTM unit for textual entailment recognition. In: COLING (2016)

Scalable Reference Genome Assembly from Compressed Pan-Genome Index with Spark

Altti Ilari Maarala[1]([✉])[iD], Ossi Arasalo[2], Daniel Valenzuela[1][iD],
Keijo Heljanko[1,3][iD], and Veli Mäkinen[1,3][iD]

[1] Department of Computer Science, University of Helsinki, Helsinki, Finland
ilari.maarala@helsinki.fi
[2] Department of Computer Science, Aalto University, Espoo, Finland
[3] Helsinki Institute for Information Technology, HIIT, Espoo, Finland

Abstract. High-throughput sequencing (HTS) technologies have enabled rapid sequencing of genomes and large-scale genome analytics with massive data sets. Traditionally, genetic variation analyses have been based on the human reference genome assembled from a relatively small human population. However, genetic variation could be discovered more comprehensively by using a collection of genomes i.e., pan-genome as a reference. The pan-genomic references can be assembled from larger populations or a specific population under study. Moreover, exploiting the pan-genomic references with current bioinformatics tools requires efficient compression and indexing methods. To be able to leverage the accumulating genomic data, the power of distributed and parallel computing has to be harnessed for the new genome analysis pipelines. We propose a scalable distributed pipeline, PanGenSpark, for compressing and indexing pan-genomes and assembling a reference genome from the pan-genomic index. We experimentally show the scalability of the PanGenSpark with human pan-genomes in a distributed Spark cluster comprising 448 cores distributed to 26 computing nodes. Assembling a consensus genome of a pan-genome including 50 human individuals was performed in 215 min and with 500 human individuals in 1468 min. The index of 1.41 TB pan-genome was compressed into a size of 164.5 GB in our experiments.

Keywords: Computational genomics · Genome assembly · Compression · Indexing · Big data · Distributed computing

1 Introduction

High-throughput sequencing (HTS) technologies have enabled rapid DNA sequencing of multiple samples collected from any organism and environment including human tissues, bacteria, fungi, plants, soil, water, and air. Next-generation sequencing (NGS) technology provides relatively cheap and rapid

S. Nepal et al. (Eds.): BIGDATA 2020, LNCS 12402, pp. 68–84, 2020.
https://doi.org/10.1007/978-3-030-59612-5_6

whole-genome sequencing enabling large-scale and profound genome analytics. As a result of advanced HTS technology, the sequencing data volumes are growing quickly and the number of assembled genomes is increasing rapidly as well.

Computational pan-genomics [1] is one of the efforts to exploit the huge amount of information from multiple genomes in comparative analysis bringing new opportunities for population genetics. Marcshall et al. pointed out that a pan-genome can present: (i) the genome of a single selected individual, (ii) a consensus drawn from an entire population, (iii) a functional genome (without disabling mutations in any gene), or (iv) a maximal genome that captures all sequences ever detected and generalized where the pan-genome can refer to any collection of genomic sequences that are analyzed jointly or is used as a reference [1].

The first human reference genome draft was published in 2004 by the Human Genome Project and it has been complemented from time to time [4]. Nowadays it is used as a comparative reference in the majority of scientific contributions in human genetic studies. Pan-genomes can represent more diverse populations without disabling any population- or individual-specific genomic regions, and thus, improve the genetic variation analysis by considering the genetic recombination and emphasizing the diversity of individuals [3,8,17–20]. Sherman and Salzberg [2] underline the importance and advantages of using pan-genomes in genetic variation studies instead of just a single reference genome. Yet, constructing a reusable reference genome from a human pan-genome requires assembling, indexing, and aligning of multiple genomes in a population, which is computationally demanding and time-consuming. Computational limits are hit in many of the processing steps: construction of pan-genome from multiple genomes, compressing and indexing the pan-genome, alignment of donor sequences to pan-genomic index, and finally assembling a pan-genomic consensus reference.

Our goal is to enable the assembling of compressed and reusable pan-genomic reference indexes that can be used directly for sequence alignment in genetic variation analyses efficiently. We focus on the scalable assembling of a reference genome from a human pan-genome as well as indexing and compressing large pan-genomes to reduce the computation time and to improve space-efficiency. PanGenSpark is a continuation for PanVC, a sequential pan-genomic variation calling pipeline with hybrid indexing presented by Valenzuela et al. in [16,30]. Here, we proceed with parallelizing the most compute-intensive phases in PanVC such as compressing and indexing of large pan-genomes and assembling the consensus genome from a pan-genome with distributed methods.

Analyzing a huge amount of genomic data is computationally intensive, and extremely so in the pan-genomic context. We propose a scalable distributed pan-genome analysis pipeline, PanGenSpark, to assemble a new consensus reference genome from a pan-genome for downstream analysis such as sequence alignment and variant calling. The pipeline implements distributed compressed indexing of the pan-genome, the read alignment, the consensus genome assembly method, and the support for legacy variant calling tools. The prototype pan-genome analysis pipeline is designed for the Apache Spark [22] framework. We demonstrate the pipeline with 500 human genomes in the Apache Spark cluster. The source code of PanGenSpark is publicly available in GitHub[1].

1.1   Related Work

The pan-genome as a concept was first presented by Tettelin et al. in [9] where a pan-genome was used for studying genetic variation in bacteria. Since then, pan-genomes have been used successfully in various studies for identifying microbial pathogens [5–7]. Sherman et al.[10] assembled a human pan-genome from NGS data of 910 African descendants and revealed 10% novel genetic material that was not found from the standard human reference genome. Mallick et al. [11] studied 300 human individuals from 142 diverse populations and found 5.8 million base pairs not presented in the human reference genome. Duan et al. analyzed 275 Chinese individuals with HUPAN [12] pipeline from NGS data where they found 29.5 million base pairs of novel sequences and 188 novel genes. Zhiqiang et al. demonstrated EUPAN [13] toolkit by analyzing the pan-genome consisting of 453 rice genomes [14]. In [15] we developed ViraPipe, a scalable pipeline for mining viral sequences from a large amount of human metagenomic samples on distributed Apache Spark cluster. ViraPipe has been used in an experiment with 768 whole-genome sequenced human samples. Most of these studies are based on the De-novo method that assemblies longer sequences, *contigs*, from short reads sequenced from donor DNA that does not map to reference genome. The contigs are used to form the pan-genome which is eventually used for analyzing novel sequences. This work instead is based on the whole-genome *re-sequencing* which differs from the De-novo based approaches in that individual genomes are assembled using a reference genome. We construct a complete pan-genome from previously assembled whole-genomes where a new consensus reference genome is assembled considering all variation between the individual genomes in the pan-genome. The assembled consensus genome enables then read alignment and variant calling with a complete pan-genome.

Hadoop-BAM [26] library has been originally developed for processing genomic data formats in parallel with Apache Hadoop[2] and Spark [22], and developed further under the Disq[3] project for even better Spark integration. It includes Input/Output interface for distributing genomics file formats into HDFS and tools e.g.., sorting, merging, and filtering of read alignments. Currently, supported genomics file formats are BAM, SAM, CRAM, FASTQ, FASTA, QSEQ, BCF, and VCF. Hadoop-BAM is already used in genome analytics frameworks and libraries such as GATK4[4], Adam[5], Halvade [27], Seal[6] and SeqPig[7].

---

[1] https://github.com/NGSeq/PanGenSpark.
[2] https://hadoop.apache.org.
[3] https://github.com/disq-bio/disq.
[4] https://gatk.broadinstitute.org.
[5] https://github.com/bigdatagenomics/adam.
[6] http://biodoop-seal.sourceforge.net.
[7] https://github.com/HadoopGenomics/SeqPig.

GATK is a software package for HTS data analysis developed by Broad Institute offering best practices variant discovering pipeline for human genomes. The current GATK4 version has been developed partly on Apache Spark for enabling distributed parallelization for rapid exploratory genomic studies. They have also established an open-source FireCloud platform for managing, sharing, and analyzing genomics data. ADAM is an Apache Spark-based genome analysis toolkit developed at UC Berkeley. ADAM includes basic tools for genomics file transformations, k-mer counting, and allele frequency computation on Apache Spark cluster. Halvade is a distributed read alignment pipeline based on the Hadoop MapReduce framework [21] for enabling more efficient variant calling with GATK. Halvade uses MapReduce for distributing BWA read alignment on read chunks against the reference genome. Seal is a software suite developed in CRS4 for processing sequencing data based on the Hadoop framework and it is written in Python. It provides basic tools for parallel and distributed read demultiplexing, read alignment, identifying duplicate reads, sorting the reads, and read quality control.

2   Methods

2.1   Distributed and Parallel Data Processing in Genomics

Traditional computational genome analysis algorithms and pipelines have been developed for sequential data processing in centralized computers, whereas the current evolution of high performance computing moves towards parallel algorithms and distributed data stores for efficient computation and analysis of massive data volumes. Moreover, current genome analysis tools and pipelines are typically developed on demand by the researchers relying on existing sequential algorithms. This has led to that pipelines are utilizing a mixture of command-line tools making them often poorly scalable, computationally inefficient, inflexible, and not easily parallelizable, especially in distributed computing clusters. Distributed and parallel computing frameworks enable scalable, reliable, efficient, and relatively low-cost computing in computing clusters. Cloud services provide infrastructures for deploying computing clusters easily and cheaper. Parallel data analysis with multiple distributed computing nodes brings huge performance advantages compared to a single computer. Computing takes place in the distributed working memory over distributed data sets by minimizing the intercommunication between nodes with optimal algorithms. This is achieved by dividing each computing task to the local parts, in which each node executes computing with local data.   Apache Spark [22] is an open-source framework developed for efficient in-memory distributed large-scale computing in computing clusters. Spark accelerates data analysis with in-memory processing where working sets of data can be reused and pipelined from one pipeline stage to another in-memory instead of using temporary files. Computing in Spark is based on Resilient distributed datasets (RDDs) [23], which are distributed and cached to the working memory of multiple computing nodes in a cluster. Each node assigns an executor for local tasks that are run in parallel on the multiple cores inside a node.

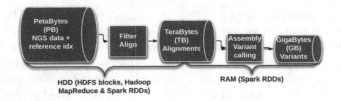

Fig. 1. Data reduction process.

Spark itself does not provide a data store, but it can retrieve and write data to Hadoop Distributed File System (HDFS). In addition, HDFS provides fault tolerance through data replication. Spark supports Scala, Java, Python, and R programming languages. Figure 1 shows a typical data reduction process of a genomics data analysis pipeline on Spark.

2.2   Distributed and Parallel Characteristics of Genomic Data

Distributed and parallel computing has not been in major focus when widely used algorithms and data models for genomics were originally designed. Data parallelism is one promising choice for parallelizing genomics pipelines without fully rewriting all of the existing algorithms. Data locality can be achieved in the nodes of the computing cluster and data processing can be done in parallel without reloading or moving any data. Raw sequencing read data can be distributed for read alignment when reference assembly methods such as BWA [24] and Bowtie [25] are used. Assembled genomes and variant data are usually parallelizable by the chromosomes and chromosomal regions, giving an opportunity to distribute input data for parallel processing stages. Existing general genomics file formats are not designed for distributed file systems and especially binary formats BAM, BCF, BED are not distributable without external tools. However, Hadoop-BAM [26] can already handle distributed BAM and BCF files on HDFS in parallel and also in-memory with Spark.

2.3   Reference Genome Assembly

Reference genome-based assembly is preceded by a read alignment process where billions of Next-generation sequencing (NGS) reads are sequenced from a donor DNA sample and aligned to a reference genome. That is, the human genome can not be sequenced as a whole with current technology. Instead, the genome is reconstructed from short fragments, called reads, which are sequenced from a donor DNA sample and aligned to a reference genome. In the pan-genomic context, the reference is a multiple sequence alignment of N sequences. The uncompressed size of a human pan-genomic reference is approximately $N \times 3$ billion bases where N is the number of haploid genomes included (human genome is diploid, having two haploids of length 3 billion bases approximately). The amount of NGS data coming from a sequencing machine depends on the size of the donor genome, the used sequencing method, and the parameters given to a sequencer. Typically the size of whole-genome sequencing read data varies from 10 to 100 GB per human genome. NGS reads have to be aligned to a reference genome for assembling consensus genome of a donor which is then used to call the variants against. The assembly process requires that every short read (tens to hundreds of base pairs long) is aligned to every position in the reference genome. This work focuses on whole-genome data where pan-genomic reference is assembled from N whole-genomes.

2.4   Compressed Indexing of Pan-Genomes

The pan-genome has to be indexed in order to perform read alignment and eventually reference genome assembly. In a pan-genomic context, the index can contain thousands of individual genome sequences. The size of the pan-genome index can be reduced hugely with compression methods such as Lempel-Ziv [28] by utilizing the characteristics of identical genome sequences between the individuals, that is, a human genome includes a large proportion of repetitive sequences which can be found from every individual. There are a few relatively fast legacy read alignment tools such as Burrows-Wheeler transformation [29] based BWA [24] and Bowtie [25]. However, the BWT based aligners use suffix array-based indexes where BWT has to permute over the whole index search space for scoring the alignment. Valenzuela et al. [30] propose a CHICO indexer based on hybrid index implementation of LZ77 variant of Lempel-Ziv algorithms for compressing the pan-genome. Hereinafter, LZ77 shall be referred to as LZ. The hybrid index separates the compression part from the indexing, where the identical parts are compressed with the LZ and the LZ compressed sequence is then indexed with the legacy Bowtie2 or BWA indexer. CHIC [31] provides also read aligner tool for aligning reads against hybrid index[8] with BWA and Bowtie2 support. The CHIC indexing with Bowtie2 was evaluated [31] on a single high-performance machine with 48 cores and 1.5 TB of main memory where they reported 35 h indexing time for 200 human genomes compressed to 180 GB index from 540 GB input data (compression ratio 3:1). Sequential PanVC[9] pipeline integrates CHIC aligner, CHICO index, and external variant calling tools such as GATK[10].

2.5   Variant Calling

Variant calling is a routine process for identifying genetic variations e.g.., Single Nucleotide Variations (SNVs) between a donor and some reference genome. Variant calling begins by aligning the NGS read sequences to a reference genome and filtering out the unaligned reads. Next, alignments are typically filtered by quality. Finally, reads are piled up over aligned reference genome base positions, and reference aligned bases are counted. This way the most probable bases in a

---

[8] https://gitlab.com/dvalenzu/CHIC.
[9] https://gitlab.com/dvalenzu/PanVC.
[10] https://github.com/broadinstitute/gatk.

donor genome covering a genomic position can be detected. Instead of using one reference genome, variant calling against a pan-genomic reference can provide more accurate information about genetic variation by aligning donor sequences to genomic positions in multiple genomes. Variant calling with a pan-genomic reference assembly can be done directly with legacy variant calling tools such as GATK's best practice pipeline.

2.6   Designing the Distributed Pipeline

Reusability of existing genomics tools is a natural starting point for designing the workflow for the pipeline as those are widely used and well known within the bioinformatics and genomics communities, and quite efficient sequential algorithms have been already developed for the most general processing phases such as read alignment. Distributed genomic data can be processed in parallel partitions at different levels (separated chromosomes, chromosomal regions, NGS read partitions) in the pipeline. We have selected to use Spark with the Hadoop Distributed File System (HDFS) in our solution as it provides a flexible framework for scalable and efficient distributed data processing, and data management. Key challenges for implementing a distributed genome analysis pipeline are; decomposing the tasks and data to partitions for parallel execution with existing genomic data formats, processing the distributed tasks in parallel with existing tools and algorithms, and piping of multiple tools and algorithms together with minimal I/O operations while maintaining load balance. Moreover, latencies for reading data from HDFS to in-memory RDDs and writing it back to HDFS have to be taken into account as bioinformatics pipelines typically process thousands of separate files as well as big files together. This sets high requirements for computing cluster's disk I/O, memory access, data warehousing, and networking performance. Figure 2 describes the architecture of the parallel pan-genomics pipeline at a high level.

2.7   Overall Pipeline Description.

The pipeline depicted in Fig. 2 consists of the following stages:

a) **Preparing the pan-genome.** The pan-genome itself is composed of multiple genomes that are aligned to a standard reference genome. Each genome in a pan-genome is assembled by applying variants from VCF files to a standard reference genome with vcf2multialign[11] tool and loading assemblies into HDFS under the same folder. When diploid genomes (e.g., humans) are used, this step generates two sequences, both haploids, per genome. The standard genome itself is applied on top of the pan-genome. If genome assemblies are already provided, the pan-genome can be constructed by simply loading all the individual genomes into HDFS under the same folder.

---

[11] https://github.com/tsnorri/vcf2multialign.

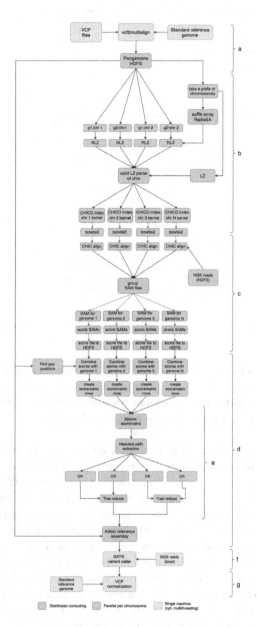

Fig. 2. The pipeline stages denoted with letters a to g are explained in the Sect. 2.7. The green stages are distributed to multiple nodes and multiple cores in parallel (with Spark). The blue stages are distributed to multiple nodes and run in parallel per chromosome (multiple cores utilized in CHIC and Bowtie). The yellow stages are run on a single node in parallel on multiple cores. (Color figure online)

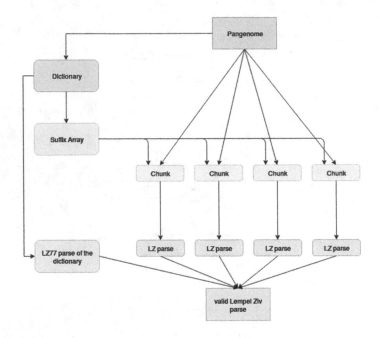

Fig. 3. Distributed Relative Lempel-Ziv compression.

b) Indexing the pan-genome. Typically, the reference index data remains a magnitudes smaller than the read data sequenced from a donor genome. In pangenomic context, the index data grows in proportion to the number of underlying individual sequences in a pan-genome where each individual sequence has to be indexed.

For reducing the compression and indexing time, we modify the CHICO indexer [30] to exploit distributed computing. CHICO supports BWA and Bowtie2 legacy indexes with the Relative Lempel-Ziv (RLZ) algorithm [32]. CHICO compresses the original pan-genome with RLZ that is eventually indexed for the read alignment purposes. The reference pan-genome is compressed to the kernel representation which reduces the repetition of similar sequences in the pan-genome. The repetitive sequences are compressed using a dictionary that is constructed from a partial pan-genome. Building the dictionary of the whole pan-genome would be too time-consuming and is not necessary due to repetitiveness, although, it can improve the compression ratio slightly. The size of the compressed kernel index depends on the number of individual genomes in the pan-genome and the similarity between the genomes.

We implement Distributed Relative Lempel-Ziv (DRLZ) compression (Fig. 3) with Spark for reducing the compression time through parallelization. The pangenome is partitioned by chromosomes of individual reference genomes and distributed to HDFS. The chromosomal chunks are RLZ compressed in parallel with Spark. RLZ uses a suffix array which is calculated from the dictionary and broadcast to Spark for distributed RLZ compression. RadixSA library is used to construct the suffix array [33]. The RLZ compressed chromosomal partitions are eventually downloaded from the HDFS to different nodes where the kernel representation is composed and indexed with CHICO in parallel per chromosome. RAM disks can be configured here for storing and accessing data in local filesystem more rapidly.

c) **Read alignment against the compressed pan-genome index.** As each chromosome is indexed in parallel, read alignment is also done in parallel per chromosome. NGS reads are loaded from the HDFS to the local filesystem of the corresponding node and aligned with CHIC aligner using Bowtie2 against a compressed index of a chromosome. Multiple sequence alignment generated gaps are stored in this step into gap position files for fixing the mapping score in the next step. Then, the mapped reads are grouped by the reference sequence where they are mapped to. After the alignment process, grouped SAM files are put to HDFS for the next step. Eventually, duplicate mapped reads are removed.

d) *Adhoc* **reference genome assembly.** After the read alignment phase, the mapped reads of each individual are scored based on the alignment information provided in the SAM and the gap position files. That is, the pan-genome index does not include gaps, and now the read mapping is fixed to correspond the gapped positions in the pan-genome [16]. The pan-genome with gaps is read from HDFS into the scoring matrix with Spark Mllib BlockMatrix class which distributes the score matrix of size $M \times N$ into S blocks, where M is the length of the individual sequence, N is the number of the individuals in the pan-genome, S is the number of partitions configured. The *adhoc* consensus genome is then assembled from the score matrix blocks by extracting the *heaviest path* (Fig. 2, e), that is simply, taking index of maximum scoring alignment per each column, mapping the index to the corresponding nucleotide in the reference genome and merging the blocks at the end.

e) **Heaviest path.** The score matrix for calculating the heaviest path is sparse; the first sequence contains most of the matches and the following genomes add only little extra information. Therefore, Apache Spark's sparse vector representation is used to achieve the best possible performance. The score matrix is transposed to find the maximum number of matches across different genomes. After this operation, we obtain the heaviest path sequence telling the row number corresponding to individual genome which has the most matches in that position. Next, the nucleotides in the corresponding positions pointed by the heaviest path are extracted from the pan-genome. To change from this number representation to corresponding DNA reference sequence, a simple logic OR operation is applied between each individual sequence in the pan-genome and the calculated heaviest path blocks (Fig. 4). After this operation has been done for each individual, a tree-reduce RDD transformation is used to combine the sequence branches into a single *adhoc* reference sequence in parallel.

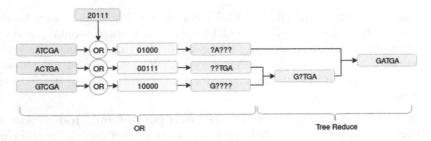

Fig. 4. Transformation from calculated heaviest path to the *adhoc* reference with three individuals.

f) Legacy variant calling from *adhoc* reference genome. After the *adhoc* reference genome assembly, variant calling is performed with the donor genome reads. The *adhoc* genome is first indexed with legacy tools such as BWA or Bowtie and donor reads are aligned to that index. Any legacy variant calling pipeline that outputs VCF format can be used to call variants from the assembled adhoc reference genome. GATK4 best practices variant calling pipeline is provided with our implementation. As an alternative, basic Samtools and Bcftools variant calling[12] method is included in our pipeline.

g) Variant normalization. If a standard reference genome is used to construct the pan-genome, the variants called from an *adhoc* reference are normalized against the standard reference genome. That is, normalization generates a consensus genome from the standard reference genome by applying variants called from the *adhoc* read alignment. In practice, normalization applies SNPs, insertions, and deletions to the reference genome positions assigned in VCF files and gap files (produced already in the indexing phase). Finally, the projection between the normalized consensus and *adhoc* reference consensus is constructed. The projection does sequence alignment between those two consensus genomes showing indels and mismatches for comparison and further analysis. The normalization and projection are done with the tools provided in the original PanVC [16].

3   Experiments

3.1   Data Preparation

We generate a pan-genome based on the human reference genome by applying heterozygous SNPs from phased haploid VCF data to GRCh37 reference thus generating two consensus genomes per individual into pan-genome. Pan-genomes are generated from the autosomes of 1000 Genomes phase 3 VCF data including 2506 individuals totaling the pan-genome size of 13.1355 TB. Read alignment and variant calling is performed with NGS read data set sequenced from donor

---

[12] http://samtools.sourceforge.net/mpileup.shtml.

HG01198 genome published by 1000 genomes project. Read data contains 9.4 million paired-end reads totalling 2 GB.

3.2 Computing Environment

The experiments are run on the Apache Spark cluster in a cloud computing environment. The cluster consists of 25 Spark worker nodes having 40 GB of RAM and 16 cores (Intel(R) Xeon(R) CPU E5-2680 v3) in each and one Spark master node having 256 GB of RAM and 48 cores. The whole cluster comprises 448 CPU cores, 1.256 TB of RAM, Infiniband 40 GB/s network, 30 TB of HDD storage space in total. The Spark cluster is deployed with Apache Spark 2.3.2 and Hadoop 3.1.0 versions on virtual machines running CentOS 7 operating system. We utilize the computing resources of the Finnish IT Center for Science (CSC) in our experiments.

3.3 Results

The scalability in terms of pan-genome size is in the main focus of our experiments. Figure 5 shows the execution time in different pipeline phases from Relative Lempel-Ziv (RLZ) compression to reference genome assembly with accumulating pan-genome size. For a baseline, a pan-genome of size 50 haploid genomes is assembled with the original PanVC on a single node. The total runtime with 50 genomes is 1624 min consisting of following execution times: 150 min for RLZ,

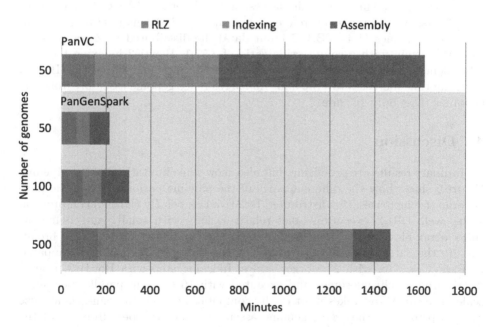

Fig. 5. Scalability of PanGenSpark with increasing pan-genome size compared to single node PanVC execution.

Table 1. Index compression ratio (CR) with N genomes in a pan-genome.

N	Uncompressed	Compressed	CR
1	2.82 GB	0.8 GB	3.53:1
50	141 GB	24.7 GB	5.71:1
100	282 GB	36.4 GB	7.74:1
500	1410 GB	164.5 GB	8.57:1

555 min for Bowtie2 indexing, and 919 min for *adhoc* reference genome assembly. Multithreaded Bowtie2 indexer is executed using 16 threads. The same multithreading configurations have been used in the distributed pipeline as well. Variant calling is not included in the results as it is not part of the distributed implementation. To note, the variant calling takes the equal amount of time with all pan-genome sizes as the *adhoc* reference genome is always the same size.

With the distributed pipeline and 50 haploid genomes (141 GB input data), the RLZ compression time is 68 min with a compression ratio of 5.71:1 shown in Table 1. Indexing time with Bowtie2 indexer is 58 min while the largest indexed kernel is 3.2 GB for chromosome 2 (chromosomes are indexed on distributed nodes in parallel using 16 threads per node). *Adhoc* reference assembly from 50 genomes takes 89 min.

With 100 haploid genomes (282 GB input data) the distributed RLZ compression time is 95 min with a compression ratio of 7.74:1. Distributed indexing time with Bowtie2 is 84 min while the largest indexed kernel is 3.8 GB for chromosome 2. Assembling the *adhoc* reference genome with 100 genomes takes 125 min. With 500 genomes (1.41 TB GB input data) the distributed RLZ compression time is 165 min with a compression ratio of 8.57:1. Bowtie2 indexing becomes a bottleneck with 500 genomes kernel (largest chromosomal kernel 16 GB) and indexing time increases to 1140 min. *Adhoc* reference genome assembly from 500 genomes takes only 163 min.

4   Discussion

Preliminary results are promising, but also show some limitations of the pipeline. Figure 5 shows how the different parts of the pipeline perform. When the pan-genome size increases, the distributed Relative Lempel-Ziv (DRLZ) compression scales well. DRLZ execution takes relatively long with small pan-genomes as suffix array files are the same size (30 GB) for all pan-genome sizes and broadcasting the data took almost 40 min. The speedup with 50 genomes compared to PanVC is 7.6x with 25 worker nodes. Indexing step with Bowtie2, that is distributed by chromosomes and executed with 16 cores in parallel per node, scales the worst and takes most of the computation when the pan-genome size grows up to 500. The *adhoc* genome assembly scales the best from all of the distributed parts. We were able to compress the index of 500 haploid genomes to the size of 58 genomes with a compression ratio of 8.57:1. For one genome the compression ratio is 3.53:1 (Table 1). The index compression ratio with sequential PanVC is equal to the distributed version as is expected. RLZ dictionary size affects the compression ratio the most (the longer it is the better the compression), and secondly the chosen maximum read sequence length parameter (the shorter it is the better the compression). We used the whole chromosomes as a dictionary for DRLZ and the maximum read length of 80 bases. Compression ratio increases in proportion to the number of genomes which is obvious due to sequence repetition.

Apache Spark is extremely good at scaling up to an unlimited number of rows, but the limiting issue here is that whole genome sequences can not be processed as one piece as the length of the string is restricted by Java *Integer.MAX_VALUE* constant which is roughly 2 billion. To improve the compression ratio we would need to use the whole genome as a dictionary part for the RLZ compression which is currently not possible due to *Integer.MAX_VALUE* issue with Spark and Hadoop frameworks. Therefore, we read the pan-genome by chromosome, compress by chromosome, and index the chromosomal kernels. BWA indexer could not handle longer than 4 GB kernel text, thus we chose to use Bowtie2. However, Bowtie2 does not scale linearly and as it can be seen from the Fig. 5: with 50 haploid genomes indexing time is 58 min whereas with 500 haploid genomes indexing time increases 20 times (the longest kernel text with 50 genomes is 3.2 GB and with 500 genomes 16 GB). However, the whole pipeline runtime increases only 6.8 times with 10 times larger pangenome showing still good scalability. Bowtie2 uses a "large" index method with 64-bit numbers when the input data size grows larger than 4 GB which seems to slow down the indexing. Bowtie2 legacy indexing time can be still decreased by harnessing more cores and memory for the indexing nodes. Tuning the cluster with all Hadoop and Spark configuration parameters turns out to be a complex task as there are plenty of those and optimal configurations vary in different processing steps. Minimizing the I/O operations is crucial when processing large data sets in the complex pipeline. Moreover, this poses challenges to the pipeline development as input data structure varies in different steps making data partitioning for parallel processing more complex. In the near future, we focus on improving the DRLZ compression as the better compression would also reduce the Bowtie2 indexing time due to shortened kernel text.

5   Conclusions

To exploit massive amounts of genomic data, it is essential to compress and store genomic data sets in an efficient and reusable form for supporting bioinformatics tools. Pan-genomic indexes and reference genomes are urgent for efficient large-scale read alignment, variant calling, and sequence matching purposes. In this work, we design a prototype pipeline, PanGenSpark, for scalable compressed indexing of pan-genomes and assembling of reference genomes from pangenomes. The PanGenSpark assembled a reference genome from a pan-genome of 50 human haploid genomes in 215 min and 500 haploid genomes in 1468 min.

The index of 1.41 TB pan-genome was compressed into a size of 164.5 GB. The experiments have been run on a distributed Spark cluster consisting of 448 cores on 26 computing nodes. Altogether, our distributed pipeline allows now assembling the pan-genomic consensus reference in a tolerable time from hundreds of human genomes in practice. Moreover, the compressed pan-genomic index can be reused for efficient NGS read alignment and sequence matching purposes as well.

Acknowledgements. The financial support of Academy of Finland project #336092 and #309048 are gratefully acknowledged. The computing capacity from CSC - IT Center for Science and the FGCI2 Academy of Finland Infrastructure project are gratefully acknowledged.

References

1. Marcshall, T., Marz, M., Abeel, T., et al.: Computational pan-genomics: status, promises and challenges. The Computational Pan-Genomics Consortium. Brief. Bioinform. (2016). https://doi.org/10.1093/bib/bbw089
2. Sherman, R.M., Salzberg, S.L.: Pan-genomics in the human genome era. Nat. Rev. Genet. **21**, 243–254 (2020). https://doi.org/10.1038/s41576-020-0210-7
3. Dilthey, A., Cox, C., Iqbal, Z., Nelson, M., McVean, G.: Improved genome inference in the MHC using a population reference graph. Nat. Genet. **47**, 682–688 (2015). https://doi.org/10.1038/ng.3257
4. Auton, A., Abecasis, G., Altshuler, D., et al.: A global reference for human genetic variation. Nature **526**, 68–74 (2015). https://doi.org/10.1038/nature15393
5. Rouli, L., Merhej, V., Fournier, P.E., Raoult, D.: The bacterial pangenome as a new tool for analysing pathogenic bacteria. New Microbes New Infect. **7**, 72–85 (2015)
6. Rasko, D.A., Rosovitz, M.J., Myers, G.S.A., et al.: The pangenome structure of Escherichia coli: comparative genomic analysis of E. coli commensal and pathogenic isolates. J. Bacteriol. **190**, 6881–6893 (2008)
7. Trost, E., Blom, J., Soares, S.C., et al.: Pangenomic study of Corynebacterium diphtheriae that provides insights into the genomic diversity of pathogenic isolates from cases of classical diphtheria, endocarditis, and pneumonia. J. Bacteriol. **194**, 3199–3215 (2012). https://doi.org/10.1128/jb.00183-12
8. Kehr, B., Helgadottir, A., Melsted, P., et al.: Diversity in non-repetitive human sequences not found in the reference genome. Nat. Genet. **49**, 588–593 (2017). https://doi.org/10.1038/ng.3801
9. Tettelin, H., Masignani, V., Cieslewicz, M.J., et al.: Genome analysis of multiple pathogenic isolates of Streptococcus agalactiae: implications for the microbial 'pan-genome'. Proc. Natl. Acad. Sci. U.S.A. **102**, 13950–13955 (2005). https://doi.org/10.1073/pnas.0506758102
10. Sherman, R.M., Forman, J., Antonescu, V., et al.: Assembly of a pan-genome from deep sequencing of 910 humans of African descent. Nat. Genet. **51**, 30–35 (2019). https://doi.org/10.1038/s41588-018-0273-y
11. Mallick, S., Li, H., Lipson, M., et al.: The Simons genome diversity project: 300 genomes from 142 diverse populations. Nature **538**, 201–206 (2016). https://doi.org/10.1038/nature18964

12. Duan, Z., Qiao, Y., Lu, J., et al.: HUPAN: a pan-genome analysis pipeline for human genomes. Genome Biol. **20**, 149 (2019). https://doi.org/10.1186/s13059-019-1751-y
13. Hu, Z., et al.: EUPAN enables pan-genome studies of a large number of eukaryotic genomes. Bioinformatics **33**(15), 2408–2409 (2017). https://doi.org/10.1093/bioinformatics/btx170
14. Zhao, Q., Feng, Q., Lu, H., et al.: Pan-genome analysis highlights the extent of genomic variation in cultivated and wild rice. Nat. Genet. **50**, 278–284 (2018). https://doi.org/10.1038/s41588-018-0041-z
15. Maarala, A.I., Bzhalava, Z., Dillner, J., Heljanko, K., Bzhalava, D.: ViraPipe: scalable parallel pipeline for viral metagenome analysis from next generation sequencing reads. Bioinformatics **34**(6), 928–935 (2018). https://doi.org/10.1093/bioinformatics/btx702
16. Valenzuela, D., Norri, T., Välimäki, N., et al.: Towards pan-genome read alignment to improve variation calling. BMC Genomics **19**, 87 (2018). https://doi.org/10.1186/s12864-018-4465-8
17. Siren, J., Välimäki, N., Mäkinen, V.: Indexing graphs for path queries with applications in genome research. IEEE/ACM Trans. Comput. Biol. Bioinform. **11**(2), 375–388 (2014). https://doi.org/10.1109/TCBB.2013.2297101
18. Huang, L., Popic, V., Batzoglou, S.: Short read alignment with populations of genomes. Bioinformatics **29**, 361–370 (2013). https://doi.org/10.1093/bioinformatics/btt215
19. Schneeberger, K., Hagmann, J., Ossowski, S., et al.: Simultaneous alignment of short reads against multiple genomes. Genome Biol. **10**, R98 (2009)
20. Paten, B., Novak, A., Haussler, D.: Mapping to a reference genome structure. ArXiv http://arxiv.org/abs/1404.5010 (2014)
21. Jeffrey, D., Sanjay, G.: MapReduce: simplified data processing on large clusters. Commun. ACM **51**, 107–113 (2008). https://doi.org/10.1145/1327452.1327492
22. Zaharia, M., Chowdhury, M., Franklin, M.J., Shenker, S., Stoica, I.: Spark: cluster computing with working sets. In Proceedings of the 2nd USENIX conference on Hot topics in cloud computing (HotCloud 2010), p. 10. USENIX Association, USA (2010)
23. Zaharia, M., Chowdhury, M., Das, T., et al.: Resilient distributed datasets: a fault-tolerant abstraction for in-memory cluster computing. In: Proceedings of the 9th USENIX conference on Networked Systems Design and Implementation (NSDI 2012), Berkeley, CA, USA, p. 2 (2012)
24. Li, H., Durbin, R.: Fast and accurate short read alignment with Burrows-Wheeler transform. Bioinformatics **25**(14), 1754–1760 (2009). https://doi.org/10.1093/bioinformatics/btp324
25. Langmead, B., Trapnell, C., Pop, M., Salzberg, S.L.: Ultrafast and memory-efficient alignment of short DNA sequences to the human genome. Genome Biol. **10**, R25 (2009). https://doi.org/10.1186/gb-2009-10-3-r25
26. Niemenmaa, M., Kallio, A., Schumacher, A., Klemelä, P., Korpelainen, E., Heljanko, K.: Hadoop-BAM: directly manipulating next generation sequencing data in the cloud. Bioinformatics **28**(6), 876–877 (2012). https://doi.org/10.1093/bioinformatics/bts054
27. Decap, D., Reumers, J., Herzeel, C., Costanza, P., Fostier, J.: Halvade: scalable sequence analysis with MapReduce. Bioinformatics **31**(15), 2482–2488 (2015)
28. Ziv, J., Lempel, A.: A universal algorithm for sequential data compression. IEEE Trans. Inf. Theory **23**(3), 337–343 (1977). https://doi.org/10.1109/TIT.1977.1055714

29. Burrows, M., Wheeler, D.J.: A block-sorting lossless data compression algorithm. Technical report 124, Palo Alto, CA, Digital Equipment Corporation (1994)
30. Valenzuela, D.: CHICO: a compressed hybrid index for repetitive collections. In: Goldberg, A.V., Kulikov, A.S. (eds.) SEA 2016. LNCS, vol. 9685, pp. 326–338. Springer, Cham (2016). https://doi.org/10.1007/978-3-319-38851-9_22
31. Valenzuela, D., Mäkinen, V.: CHIC: a short read aligner for pan-genomic references. bioRxiv 178129 (2017). https://doi.org/10.1101/178129
32. Hoobin, C., Puglisi, S.J., Zobel, J.: Relative Lempel-Ziv factorization for efficient storage and retrieval of web collections. Proc. VLDB Endow. $5(3)$, 265–273 (2011). https://doi.org/10.14778/2078331.2078341
33. Rajasekaran, S., Nicolae, M.: An elegant algorithm for the construction of suffix arrays. J. Discrete Algorithms 27, 21–28 (2014). https://doi.org/10.1016/j.jda.2014.03.001

A Web Application for Feral Cat Recognition Through Deep Learning

Jingling Zhou, Shiyu Wang, Yunxue Chen,
and Richard O. Sinnott$^{(\boxtimes)}$ iD

School of Computing and Information Systems, University of Melbourne,
Melbourne, Australia
rsinnott@unimelb.edu.au

Abstract. Deep learning has gained much attention and been applied in many different fields. In this paper, we present a web application developed to identify and detect the number of *distinct* feral cats of Australia using deep learning algorithms targeted to data captured from a set of remote sensing cameras. Feral cat recognition is an especially challenging application of deep learning since the cats are often similar and, in some cases, differ only in very small patterns on their fur. Given the automated, sensor-based image capture from remote cameras, further challenges relate to the limited number of images available. To tackle this, we train four neural network models to distinguish 75 classes (i.e. distinct feral cats) using 30 to 80 images for each class. Based on a range of evaluation metrics, we select Mask R-CNN model with ImageNet pre-trained weights augmented with the ResNet-50 network as the basis for the web application. Using images of cats from cameras in five different forests around Victoria, we achieved an average accuracy of identification of individual (distinct) cats of 89.4% with a maximum accuracy of 96.3%. This work is used to support ecologists in the School of Biosciences at the University of Melbourne.

Keywords: Deep learning · R-CNN · Faster R-CNN · Mask R-CNN · Keras · Feral cat recognition

1 Introduction

In the last 200 years, Australia has had more mammal extinctions than any other country and native species are continuing to decrease at an exponential rate [1]. The major cause of this is humans, either directly through deforestation or indirectly through the release of non-native species/predators. Feral cats are one example of a predator that has caused havoc with the indigenous species of Australia. These cats have spread across the entire country but are especially prevalent in rural and wooded areas. Despite their relative abundance, it is not known how many cats there are and how this varies spatially. It is therefore important to better understand feral cat density and their movement patterns. Remote sensing cameras with cat traps, e.g. food accessible to cats that are used to trigger image capture through motion detection, are often used to obtain images of feral cats. However distinguishing individual (different) feral cats manually by visual inspection is extremely time-consuming and tedious. The

S. Nepal et al. (Eds.): BIGDATA 2020, LNCS 12402, pp. 85–100, 2020.
https://doi.org/10.1007/978-3-030-59612-5_7

goal of this work is to explore how deep learning/convolutional neural network (CNN) models [2] can be used to tackle this problem. These models are benchmarked for their accuracy and a web application established to allow biologists automatically identify individual feral cats from the captured images. This paper describes these models and the results of their application, as well as the web application to support the domain scientists.

The rest of this paper is structured as follows. Section 2 describes the background to deep learning and the models that are available. Given the importance of data for deep learning, Sect. 3 focuses on the data capture process and how sufficient, high quality images were obtained. Section 5 focuses on the results of the model application and Sect. 6 providing an overview of the web application to support the work. Finally, Sect. 7 draws conclusions on the work as a whole and identifies potential areas of future improvement.

2 Background to Deep Learning and Related Work

Deep learning is one branch of machine learning based on artificial neural networks. Deep learning is now widely used and explored by many researchers and companies alike in areas such as computer vision, image recognition, drug design, machine translation and more. Compared to traditional machine learning methods, which need manual feature selection and adjustment. Deep learning models are based on artificial neural networks using several layers, where each layer is able to transform input into a more composite and abstract representation. Through this approach, deep learning is able to find more accurate and abstract features than possible through manual feature selection [3]. One common application of deep learning is *object detection* and *classification* for computer vision and image processing. *Object detection* is used where a user inputs an image, the model will detect particular objects in the image, e.g. *is there a cat in the image?* If so, it will return the locations of potential cats in the image as a bounding box or a mask surrounding the object. *Object classification* is used to classify the object, e.g. *what is the particular breed of cat.* In this paper, we apply deep learning algorithms to detect whether a feral cat is in an image and return its label with a bounding box, which is then used to classify the particular feral cat, i.e. is it the same cat identified previously or a new cat.

For object detection and classification, there are many different deep learning models that have been put forward. These models have their own innovations and processes used to support object detection and classification. The performance and accuracy of these models can vary widely, even with the same dataset as identified by Wang et al. [4]. They identified that Faster Region Proposal Convolutional Neural Network (Faster R-CNN) had better performance than other contemporary models in terms of accuracy, sensitivity and specificity. They identified that other deep learning models such as Single Snapshot Detection (SSD) and You Only Look Once (YOLO) were sub-optimal when dealing with low quality and especially with low quantities of data, however they were advantageous with regards to speed of classification, since they are based on a single phase approach unlike R-CNN which proposes regions that are then used for classification.

There are several core aspects typically associated with deep learning: *transfer learning, feature extraction* and the *deep learning framework* itself. Transfer learning is a common machine learning technique. Traditionally, the accuracy of machine learning methods is largely dependent on access to and use of high-quality labeled data. For some scenarios, transfer learning can extract pre-trained knowledge from other sources and apply it to a given target domain [4]. Since the number of images of feral cats in this work is (relatively) limited, it is essential to use transfer learning to improve the model performance and reduce the training overheads. There are many popular pre-trained datasets available. Microsoft COCO [5] and Google ImageNet [6] are two large scale examples. The COCO dataset is mainly used for object segmentation and recognition. It contains more than 330 k images including 220 k labeled images. 80 object categories contained in this dataset include the target object 'cat'. It is noted that the images in COCO often have complicated backgrounds and may contain many object targets in one image, i.e. not just images of cats. The size of the target object may also be a relatively small part of the image. Google ImageNet is another pre-trained dataset commonly used for computer vision. It contains 14,197 k images with 1034 k images including a bounding box. However, neither of these resources has a focus on feral cats.

There are several methods that have been applied for object detection including Faster R-CNN [7], Mask R-CNN [8], YOLOv3 [9] and SSD [10]. Faster R-CNN and Mask R-CNN grew out of R-CNN as originally proposed by Girshick et al. [11]. Fast R-CNN was proposed in [5] to reduce the time related to region computation. In the Fast R-CNN architecture, a region of interest (RoI) pooling layer was applied for extracting fix-length feature vectors. Each output of the RoI layer is fed to fully connected layers as inputs that branch into two output vectors. In the final state, Fast R-CNN replaces the SVM with a *softmax* layer for the classification. It uses the output vectors to predict the observed object. However, the region proposal step still consumes considerable time and computational resources using the selective search algorithm as the detection network. Therefore, Faster R-CNN introduced Region Proposal Networks (RPNs) to replace the selective search.

Mask RCNN was published by He et al. [8] as a new framework for object instance segmentation where each RoI extends Faster R-CNN. It is used to predict an object mask based on existing bounding boxes. The result of Mask R-CNN not only shows the bounding box and class label, but also an object mask suitable for image recognition problems. Mask R-CNN addresses the misalignment of Faster R-CNN, which is based on network inputs and outputs following a pixel-to-pixel alignment [8]. A new simple, quantization-free layer (*RoIAlign*) is also added. With this layer, the whole instance segmentation process can preserve correct spatial information up until the final result presentation. As a result, the accuracy of the mask location is improved significantly.

The *RoIAlign* layer is designed to align the extracted features with the input without the need for strict quantization of the *RoIPool*. The quantization in the *RoIPool* does not impact label classification, but it can lead to a misalignment between the extracted features and the RoI. To avoid this, a new bilinear interpolation layer is used to replace the RoI boundaries quantization. As a result, Mask R-CNN has a performance improvement for object detection and is fast to train.

ResNet is a residual learning framework suitable for training of potentially deep networks [12]. It reconstructs the layers to learn residual functions based on the layer inputs. Furthermore, these residual networks can be easily optimized to improve the accuracy gained from increasing the depth, which addresses the degradation problem. It also addresses the problem of vanishing gradients since the gradient signal can propagate back to previous layers with the help of shortcut connections, which allows more layers to be built into the network. *ResNet* performs well for detection and localization using ImageNet and for detection and segmentation for COCO, hence in this work we use *ResNet* with different numbers of layers (*50-layer ResNet* and *101-layer ResNet*). The different number of layers has implications for accuracy and also for the performance of the models. As with all deep learning approaches, the models themselves are hugely dependent on the quality and quantity of the input data.

3   Dataset

All of the images used in this work are based on the deployment and use of heat-in-motion sensing camera-traps. These cameras are triggered when an animal moves in front of them. Feral cats have naturally uniquely markings, e.g. individual spots and stripe patterns, they can in principle be distinguished. The photos themselves were taken from two main sites around rural Victoria: the Otway Ranges and Glenelg Region in south-west Victoria. Camera-traps were placed in a range of grids across these sites as shown by the dotted white regions in Fig. 1. Each grid had 70–110 cameras that were operational for 2–3 months. In the Otway Ranges, there were two grids in total and feral cats could appear in either grid since the grids were quite close together. In the Glenelg Region, there were four grids which were far away from one another, hence individual cats could not appear in the different grids.

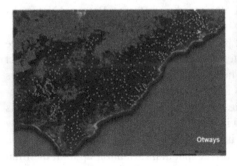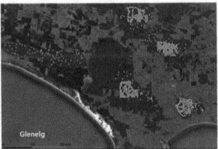

Fig. 1. Otway and Glenelg regions around Melbourne from satellite photographs

Since the images come from photos of potentially moving animals, the quality of many images can often be quite low. For example, as shown in Fig. 2, image 1 is too blurry to distinguish the pattern of the cat, and image 2 only contains the tail of the cat. In addition, the two cats overlap in image 3 which makes it difficult to separate the pattern of each cat, whilst image 4 is over-exposed and hence the features of the cat cannot be detected. To address this, we selected images of high quality showing different sides of each individual cat. Individual cats with less than 5 high quality photos were combined them into an "other" class.

Fig. 2. Examples of feral cat images with low quality

After image selection, various preprocessing steps were necessary before the data could be used for training the models. Since there were less than 10 selected images for some feral cats and more than 50 selected images for other cats, the dataset was unbalanced. This can lead to poor classifications for cats, especially those with fewer images. Data augmentation is commonly used to expand and balance datasets. Data augmentation can involve many methods. *Rotation* and *horizontal* flipping were used in this work. By rotating images through a suitable angle, it is possible to get additional images Fig. 3 shows the rotation of one image by 0–60 degrees. It is noted that the rotation does not impact the accuracy of the model in detecting and classifying the cat.

Horizontal flipping involves mirroring the image. Figure 4 shows the result of horizontal flipping of one image.

For object detection, resizing an image will not affect the result since the bounding box information is recorded in the annotation files. In the training phase, the information within the bounding box is extracted. However, to support the training, multiple images per batch were used in the training phase to increase the training efficiency. All images were resized to 1024×1024 pixels.

After pre-processing, the total number of images obtained was 4,226 and the number of classes (individual feral cats) was 93. Since the datasets were relatively small but the number of classes relatively large, the accuracy of models may well be low. To make the model more accurate, other information was incorporated into the work. For example, since the four grids in the Glenelg Region were separated, it was

Fig. 3. The result of image manipulation before and after rotation

Fig. 4. The result of an image before and after horizontal flipping

known that individual cats could not appear in more than one grid. As a result, we treated the four grids in Glenelg Region (referred to as *Annya, Hotspur, Mt_Clay* and *Cobbob*) as four separate datasets. We also treat the Otway Ranges as an independent dataset. In addition, we combine individual cats with only a few images of high quality together as a single "other" class. With this approach, the number of classes was reduced to 75. The distributions of images and classes for each dataset is shown in Fig. 5.

Note that the cats were all assigned representative names based on their markings by the ecologists that were originally tasked with (manually) counting the number of unique feral cats. The Python tool '*labelImg*' was used to manually annotate images, including creating labels and bounding boxes information. These were saved as.*xml* files. Once created, the datasets were shuffled and divided into a *training set* and *validation set* in the ratio 4:1.

4   Model Construction and Comparison

In this section, we illustrate the process of model construction and how we select the final model by comparing different methodologies. By combining the frameworks, feature extractors and transfer learning models introduced above, we built four object detection models as alternatives, namely: *Mask R-CNN + ResNet101 + COCO*; *Mask*

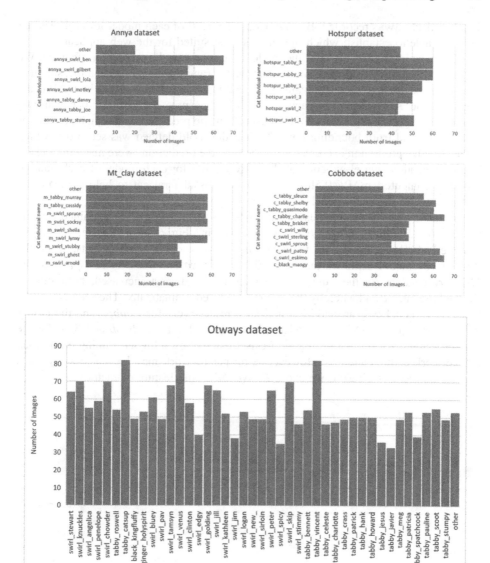

Fig. 5. The number of feral cat images in the different regions

R-CNN + ResNet50 + COCO; Faster R-CNN + ResNet50 + ImageNet, and Mask R-CNN + ResNet50 + ImageNet.

During the training phase, we used the COCO and ImageNet pre-trained models to support transfer learning. Then, we retrained the models based on the feral cat datasets. It should be noted that the number of iterations (epochs) was set to 40 and the training steps of each epoch was set to 100. The mean Average Precision (mAP) of each model

92   J. Zhou et al.

for each dataset was recorded. By comparing the *mAP* of each model for each dataset, we were able to establish which model was most suited for the web application.

The most important aspect of the model in this work is the prediction accuracy. We chose mAP@0.5IoU as the evaluation metric. Average precision(AP) is a measure that computes the average precision value for recall value between 0 to 1. The mAP is the mean of the AP values for each class. IoU measures the overlap between two boundaries. Specifically, we use it to measure how the predicted bounding box overlaps with the ground truth bounding box. The IoU threshold defined was set to 0.5, which means that the prediction is correct if IoU >0.5 in the predictions were made for a class. In the evaluation phase, we computed mAP@0.5IoU for each model of each dataset. We also computed mAP@0.5IoU for each individual feral cat to help compare the performance of the final models.

Initially, we built two models for Mask-RCNN and used COCO pre-trained weights for *ResNet-50* and *ResNet-101*. As shown in Fig. 6, the ResNet-50 model achieved >90% accuracy with the *Hotspur, Annya* and *Mt_Clay* datasets, while the ResNet-101 model achieved just over 80% for *Annya* and 70% for *Hotspur* and the *Mt_Clay* dataset. As the number of classes in the dataset increased, both models showed the same trend whereby the accuracy reduced dramatically. The accuracy of the ResNet-50 model was around 75% and 39% respectively for *Cobbob* and the *Otway* dataset, while the *ResNet-101* model only reached 61% for *Cobbob* and 34% for the *Otway* Ranges. As discussed in [12], ResNet can exhibit a higher training error when the network depth increases. However, a deeper network may be unnecessary for a smaller dataset. As a result, we chose *ResNet-50* as the feature extractor.

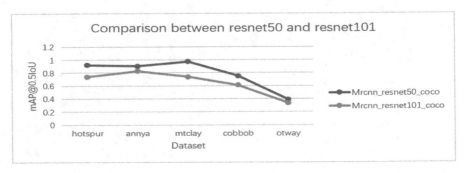

Fig. 6. Comparison between ResNet-50 and ResNet-101

We also compared two models of Mask-RCNN built on ResNet-50. One model used ImageNet pre-trained weights, whilst the other used MS COCO pre-trained weights. Figure 7 shows that the performance of the two models has no great difference for the *Hotspur, Annya* and *Mt_Clay* dataset, whilst the accuracy of both models decreases for the *Cobbob* and *Otway* dataset. However, the accuracy of the ImageNet model is about 20% and 36% greater than the COCO model in the *Cobbob* and *Otways Region* dataset respectively.

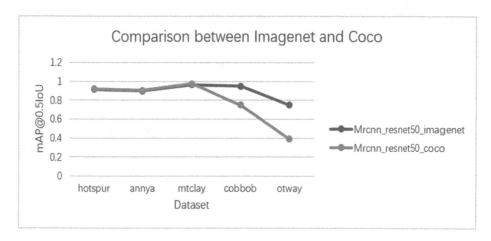

Fig. 7. Comparison between ImageNet and COCO data sets

As mentioned, the number of categories in the MS COCO dataset is much smaller than ImageNet, but the average categories and instances per image are higher than ImageNet. In addition, the percentage of small objects in the MS COCO dataset is much higher than ImageNet, so MS COCO performs better on small object detection. Since the structure of the datasets are closer to ImageNet and there is typically only one cat in each image and the number of classes is high, the cat object in an image is relatively large, hence ImageNet performs better than MS COCO overall. As a result, we chose ImageNet as the pre-trained model for the web application.

We compared two models built on ResNet50 with ImageNet pre-trained weights. One model used the Faster R-CNN framework and the other used Mask R-CNN. According to Fig. 8, for the *Hotspur, Annya, Mt_Clay* and *Cobbob* dataset, the Mask R-CNN model had a better overall performance than Faster R-CNN. For the *Cobbob* dataset, the Mask R-CNN model achieved 94.7% accuracy, while Faster R-CNN only achieved 76.9% accuracy. However, Fig. 8 also shows that the accuracy of the Faster R-CNN model is slightly better than the Mask R-CNN for the *Otways* dataset. Since the overall performance of the Mask R-CNN model is better than the Faster R-CNN model, we chose the Mask R-CNN as the framework for the final model.

In summary, we chose ImageNet as the pre-trained model, Resnet50 as the feature extractor and Mask-RCNN as the framework to build the final model. The result and analysis of the final model is discussed in Sect. 5.

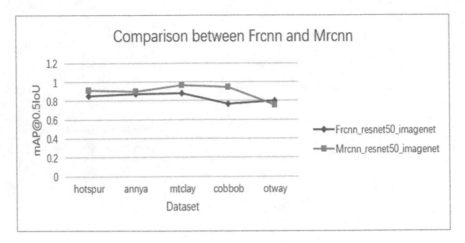

Fig. 8. Comparison between Faster R-CNN and Mask R-CNN

5 Results and Analysis

The work leveraged the Google Colab research environment. The time for training *Hotspur, Annya, Mt_Clay* and *Cobbob* datasets was around 4 h per model, and the training time for the *Otway* dataset was around 6.5 h.

The mean average precision with 0.5 Intersection over Union (mAp@0.5IoU) for all datasets is shown in Fig. 9. As seen, the *Mount Clay* forest reached a maximum accuracy of 0.963 compared to the other datasets, while the *Otway* forest dataset had the lowest mAPat 0.75. By analyzing the results from the four small datasets *Hotspur, Annya, Mt_Clay* and *Cobbob*, it was found that the training model with small datasets had better performance than large datasets like Otway. The average mAP value of the four datasets was 0.93. The difference in results is due to the fact that the Otway region has more classes than others, but the number of images of each cat was relatively low.

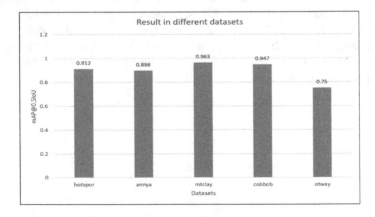

Fig. 9. The results of the different datasets

As a result, the model may not learn enough features from objects in the Otway region. Furthermore, several cats in the Otway region had almost the exact same pattern leading to erroneous classifications. A further factor was that there was a lot of low-quality cat images in the Otway region compared to the other datasets.

After analyzing the mAP value of each dataset, the mAP for each cat was calculated. As shown in Fig. 10, the distribution of individual mAP value in *Annya* and *Cobbob* was similar with mAP values of all cats above 0.8, except for the cats labelled as *swirl_1* and *swirl_3* in the *Hotspur* region, which had a mAP of 0.75 and 0.78 respectively.

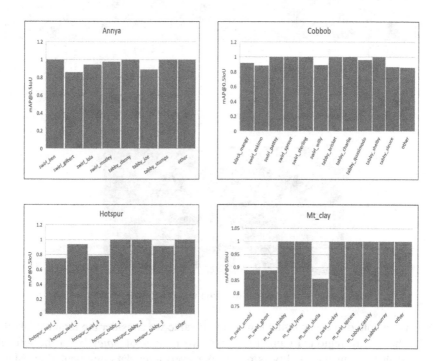

Fig. 10. The accuracy of individual feral cats in the smaller datasets

Figure 11 shows the mAP value of each individual cat in the Otway Ranges. It is clear to see that there is an unbalanced mAP value distribution in this dataset. Some cats can be fully recognized with mAP = 1, while some cats have relatively low mAP value, e.g. *Otway_swirl_spicy* has a mAP of 0.214. As an example of the highest mAP value, the cat labelled *stubby* in the *Mt_Clay* dataset is shown in Fig. 12. From Fig. 12, we can see that this image has very high quality and the boundary between the cat and background is very clear. The pattern of *stubby* is also easy to distinguish.

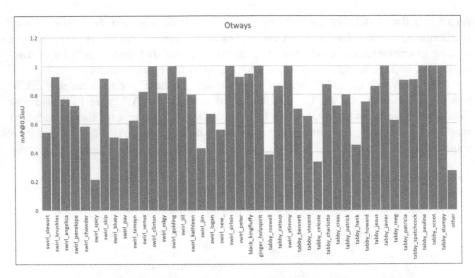

Fig. 11. The accuracy of individual feral cats in the Otway region

Fig. 12. The Feral cat (*Stubby*) in the Mt_Clay region

As an example of a cat with the lowest mAP value we consider the feral cat labelled *Spicy* in the Otway region. Figure 13 shows two images of *Spicy*, with the left one illustrating that the image quality is quite poor even after data augmentation. Meanwhile in the right figure, this cat is mixed with the background further challenging the cat identification. In addition, some cat patterns are very similar to each other in the Otway region. Figure 14 shows the patterns of two cats: *Spicy* and *Skip* in the Otway region, where it can be seen that it is non-trivial to distinguish them.

Fig. 13. The Feral cat (*Spicy*) in Otway

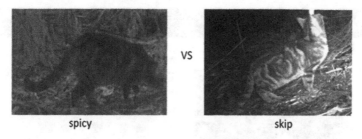

VS

spicy   skip

Fig. 14. Comparison between the feral cats labelled as *Spicy* and *Skip*

6 Web Application

The deep learning solution was delivered as a web application that was deployed on the National eResearch Collaboration Tools and Resources project (Nectar) Research Cloud (www.nectar.org.au). Nectar provides free, scalable computational power to all Australian researchers. The Nectar Research Cloud also provides a dedicated availability zone for The University of Melbourne.

In the development of the web application, the Mask R-CNN model for feral cat identification and detection was used. The web application automatically loads all models with their corresponding configuration. Each model has its own Tensor Flow Graph session instance which is loaded in the model. There are three main pages in the website: a home page, a demonstration page and a resultspage.

The design philosophy of the website had two simple tenets, a neat and simple interface as shown in Fig. 15. The system provides five trained models for each forest, where the location of the forest is indicated on the map. Once a user chooses a forest, the marker will display on the corresponding position on the map.

Fig. 15. User interface for forest selection

The results page is shown in Fig. 16. The right-hand side of the page includes several parts: the feral cat label and calculated confidence, reference photos and the map.

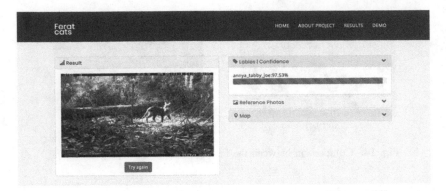

Fig. 16. Results page

There are several usage scenarios considered in the web application: identifying a cat from a given forest and identifying a cat with unknown origin. After image uploading, the system will return the original image with predicted bounding box and the results as shown in Fig. 17.

Fig. 17. Original image (left) and image with bounding box (right)

Figure 18 shows the result after a user selects the specific forest (*Mount Clay*) for the system to load model and uploads an image of a cat from that forest (a cat labelled as *swirl_ghost*). As seen, the confidence level achieves over 98%.

Fig. 18. The result of detecting the cat from its own forest

In Fig. 19, shows the result of the system detecting a cat in a forest where it does not belong. In this case the cat "*swirl_eskimo*" and the Cobbob forest are used as an example – as seen the classification confidence is greatly reduced (28%). By comparing the reference photos with the uploaded photo, it is easy to identify that the patterns of these cats have something in common, both of them are physically similar and they all have the circle pattern in their right side.

Fig. 19. Classifying a cat in a forest where it does not belong

When there is no forest chosen and the user uploads a photo, the system will consider all forests. In this case, the result is shown in Fig. 20. As seen, the highest object detection and classification result is for the feral cat*swirl_ben* from the *Annya* forest with a confidence of over 96%.

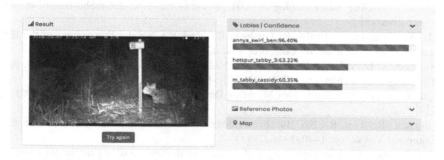

Fig. 20. The result of detecting a feral cat of unknown origin

7   Conclusions

Identifying feral cats individually is time-consuming and tedious. In this work, we explored deep learning models to help identify cat individuals automatically. During the data pre-processing, we select photos of high quality and use data augmentation to expand and balance each dataset. We compared the performance of ResNet-50 and ResNet-101, COCO and ImageNet as well as Mask-RCNN and Faster-RCNN. Ultimately, we chose ResNet-50 for feature extraction, ImageNet as the pre-trained model and Mask-RCNN as the framework to build the final model. In the evaluation, we used mAP with IoU = 0.5 as the evaluation metric. We obtained a mean average precision of 91.22%, 89.82%, 96.35%, 94.74% and 75.03% for the *Hotspur, Annya, Mt_Clay, Cobbob* and *Otway* forests respectively. A web application was developed for users to detect and recognize feral cats using the models. This supported a range of key scenarios required by the ecologists at the University of Melbourne.

The source code for this work is at: https://github.com/PhoebeYunxue/Project-Freal-Cats and a video showing the functionality available at: https://youtu.be/NC3fTro64FY.

References

1. Weinstein, B., et al.: Geography of current and future global mammal extinction risk. PLoS ONE **12**(11), e0186934 (2017)
2. Liang, M., Hu, X.: Recurrent convolutional neural network for object recognition. In: Proceedings of the IEEE Conference on Computer Vision and Pattern Recognition, pp. 3367–3375 (2015)
3. Bengio, Y., Lamblin, P., Popovici, D., Larochelle, H.: Greedy layer-wise training of deep networks. In: Advances in Neural Information Processing Systems, pp. 153–160 (2007)
4. Wang, H., Yu, Y., Cai, Y., Chen, X., Chen, L., Liu, Q.: A comparative study of state-of-the-art deep learning algorithms for vehicle detection. IEEE Intell. Transp. Syst. Mag. **11**(2), 82–95 (2019)
5. Lin, T.-Y., et al.: Microsoft COCO: common objects in context. In: Fleet, D., Pajdla, T., Schiele, B., Tuytelaars, T. (eds.) ECCV 2014. LNCS, vol. 8693, pp. 740–755. Springer, Cham (2014). https://doi.org/10.1007/978-3-319-10602-1_48
6. Fei-Fei, L.: ImageNet: crowdsourcing, benchmarking & other cool things. In: CMU VASC Seminar, vol. 16, pp. 18–25, March 2010
7. Javier, R.: Faster R-CNN: Down the rabbit hole of modern object detection, 18 January 2018. https://tryolabs.com/blog/2018/01/18/faster-r-cnn-down-the-rabbit-hole-of-modern-object-detection/
8. He, K., Gkioxari, G., Dollár, P., Girshick, R.: Mask R-CNN. In: Proceedings of the IEEE International Conference on Computer Vision, pp. 2961–2969 (2017)
9. Redmon, J., Divvala, S., Girshick, R., Farhadi, A.: You Only Look Once: Unified, Real-Time Object Detection, arXiv:1506.02640 [cs], June 2015
10. Liu, W., et al.: SSD: Single Shot MultiBox Detector, arXiv:1512.02325 [cs], vol. 9905, pp. 21–37 (2016)
11. Girshick, R.: Fast R-CNN. In: Proceedings of the IEEE International Conference on Computer Vision, pp. 1440–1448 (2015)
12. Wu, Z., Shen, C., Van Den Hengel, A.: Wider or deeper: revisiting the resnet model for visual recognition. Pattern Recogn. **90**, 119–133 (2019)

MCF: Towards Window-Based Multiple Cuckoo Filter in Stream Computing

Ziyue Hu[1,2], Menglu Wu[1,2], Xiaopeng Fan[1], Yang Wang[1(✉)],
and Chengzhong Xu[3]

[1] Shenzhen Institutes of Advanced Technology, Chinese Academy of Sciences,
Beijing, China
yang.wang1@siat.ac.cn
[2] University of Chinese Academy of Sciences, Beijing, China
[3] State Key Lab of Iotsc, University of Macau, Macau, China

Abstract. In this paper, we present a new stream-oriented filter, named
Multiple Cuckoo Filter (MCF), to support concise membership queries
on multiple data streams. MCF is composed of a group of standard
cuckoo filters, in which the membership query are decomposed into a
set of single queries. MCF allows each cuckoo filter to be configured
dynamically by changing the size of sliding window. It stores elements'
fingerprint, instead of elements themselves, here fingerprint is a bit string
which is determined by a hash function. MCF can check whether a given
item exists in multiple data streams simultaneously. It is proved that
MCF outperforms better than traditional cuckoo filter on false positive
in theory. Experiments demonstrate that the query time of MCF grows
linearly with the growth of number of cuckoo filters, decreases gradually
with the growth of sliding window number, and increases with the growth
of total elements.

1 Introduction

With the rapid development of mobile Internet, Web2.0, and smart devices etc.,
the amount of data produced by human beings are growing dramatically. The
key characteristics of the "Big Data" start to unveil, which are variety, velocity and volume [13]. The multi-dimensional data increasingly brings challenges
to our information systems. So it's very important to have better methods to
calculate, search and analyze timely during massive data processing. Unlike one
data stream, multiple data streams enable systems to support a large amount of
comprehensive calculation and strike a balance of performance and efficiency.

The data stream filtering problem can be defined as determining a set of
tuples in data stream that meet certain criteria quickly. The multidimensional
element querying is a method which can judge whether the specified object exists
in a target data set. It queries among the whole dimension or part in the data set
and returns "True" or "False". The approximate matching algorithm, also known
as the approximate hash algorithm or fuzzy hash algorithm, can recognize the

© Springer Nature Switzerland AG 2020
S. Nepal et al. (Eds.): BIGDATA 2020, LNCS 12402, pp. 101–115, 2020.
https://doi.org/10.1007/978-3-030-59612-5_8

similarity between digital objects. It is widely used in algorithms such as Ssdeep [12], Sdhash [18], Mrsh-v2 [6], Bloom filter [4], Cuckoo filter [9], etc.

Bloom filter is a concise information representation scheme, which uses a bit string of vector on behalf of a set and can effectively support the lookup for elements [4]. It is a kind of data structure which can represent sets and support queries. Although its query algorithm has certain false positive rate, it takes constant lookup time and has compact storage space. Therefore, it still has high application value. The bloom filter algorithm is extensively studied with respect to the representation and query of uni-dimensional elements such as Standard Bloom filters [4], Counting Bloom filters [10], Blocked Bloom filters [16], d-left Counting Bloom filters [5], Quotient filters [3]. These algorithms discuss and optimize the design of Bloom filters from different perspectives to satisfy different needs of practical applications.

However, traditional bloom filter cannot dynamically delete existing elements, unless reconstructing the whole filter or introducing higher false positive rate. Some variants of bloom filter, such as Counting Bloom filters and d-left Counting Bloom filters, can support dynamic deletion, but both of them have increasing space or performance overhead. To reduce the overhead, Cuckoo filter [9] is put forward in 2014. It uses bit string generated by hash function, instead of key/value pair, to store fingerprints. Thus it remarkably improves the space utilization of the filter. Cuckoo filter not only supports deletion but also can guarantee the space efficiency when the false positive rate is less than 3%.

In the era of big data, there are rich and diverse sources of information, but each data stream may only provide part of the information and cannot meet the increasingly diverse application needs. There is often a certain degree of correlation between information from multiple different data streams. Combining data from multiple streams, integrating and obtaining comprehensive information become an imperative trend. Since the Cuckoo filter has not been applied to the multidimensional element membership queries currently, this paper analyses the representation and query of Cuckoo filters and then presents Multiple Cuckoo Filter (MCF) which can be applied to dynamic data stream. Based on the Cuckoo filter, MCF divides the query of multiple data streams into some logical single data stream. The number of logical single data streams is determined by the number of data streams. Each data stream corresponds respectively to a standard cuckoo filter. In this paper, the main purpose of MCF is to study whether there exists objects which satisfies the specified relationship simultaneously between different data streams and then return the results. Further, it can form a basis for building multiple data streams' correlations.

2   Related Work

In this section, we briefly introduce some related work. We list some existing filters, including both uni-dimension and multi-dimension Bloom filters. We mainly introduce cuckoo filter data, another compressed version of Cuckoo hashing table, which is highly related to our design.

2.1   Uni-Dimensional Filters

Standard bloom filter [4] has a m-bit hash table and k hash functions. When insert elements, each element x_i will get k hash values $h_j(x_i)$ $(1 \leq j \leq k)$, mapped by k independent hash functions $(h_1, h_2, \ldots, h_k)$. Then set the value $BF[h_j(x_i)]$ as 1. When query an element x, calculate k hash values with k hash functions firstly, then check if all k locations are 1. Once any location is 0, x is definitely not in the data set. If all are set by 1, then element x has a high chance to be in the data set because of false positive. The standard bloom filter uses a bit string to express a data set. We can lookup elements but cannot delete any element in standard bloom filter.

The *Counting bloom filter* (CBF) [10] is based on Standard Bloom Filter. It implements the deletion by adding a counter for each bit. When a new element needs to be inserted, we not only need to set the corresponding bit to 1, but also increase the counter. If we want to delete an element, we decrease the k corresponding counters by 1 if the counters are larger than 0. In this way, it is able achieve deletion but has more space overhead.

The *blocked Bloom filter* [16] is composed of a series of relatively small Bloom Filters, each element is located by hash partition and respectively mapped into a small Bloom Filter. In order to obtain the best performance, these small filters are stored in a cache-line-aligned fashion. Therefore, it can provide better spatial locality. Every query at most results in a cache miss to load the Bloom filter. However Blocked Bloom filter does not support the deletion, and due to the unbalanced load of the small bloom filter, it causes a higher false positive rate.

The *Quotient Filter* [3] stores the fingerprints and supports the deletion. It uses linear detection as a conflict resolution strategy. Elements having the same quotient are stored in a bucket continuously. If hash collision occurs, the element needs to be stored forward if an empty bucket is found. Quotient filter needs additional metadata to encode each entry. Therefore it occupies 10% to 25% more space than Standard Bloom filter. When the occupancy ratio of hash table is more than 75%, its performance significantly declines.

2.2   Multi-dimensional Filters

The typical solutions for multidimensional element query include *table index* [7,17], *tree index* [14,22] and *Bloom filter index* structure. Among them, the table index stores multi-dimensional elements in a table, but it has a drawback of occupying too much memory. The index needs to be dynamically maintained while adding, deleting or modifying items, which reduces the speed of data maintenance. The tree index structure supports the multi-dimensional element queries, but it is difficult to balance the load of the heterogeneous datasets' structure. And the complexity of application integration is huge. There are some typical data structures of multi-dimensional expansion of Bloom filter.

Multiple-Dimension Standard Bloom Filter (MDBF) [11] consists of a number of standard Bloom filter. The number of Bloom filters is the same with dimensions. An independent Bloom filter is built for each attribute. The query of a multidimensional element is directly decomposed into a set of querys in single attribute.

The *Combined Multi-Dimension Bloom Filter* (CMDBF) [21] bases on the Multi-Dimensional Bloom filter. It consists of two-part filters. The first part is the subset of attributes represented by the standard Bloom filter, using the same mechanism as the multi-dimensional Bloom filter. The second part is a combine Bloom filter that is used to unite all attribute domains of the element and the query confirmation. Every query needs two steps. The first step is to check whether each single-dimension attribute is in a single-dimensional attribute subset. If the attribute is found to have been mapped to the single-dimensional Bloom filter, go to the second step to check whether all attributes can be found in Multi-Dimensional Bloom filter. If all attributes of the element are found, the element belongs to the set.

Bloom filter Matrix (BFM)[20] stores a bit vector in each dimension to construct a Bloom filter. It builds a bit matrix based on independent attributes' Cartesian product for the full attribute combination query. In this way, it fundamentally eliminates the combined error rate. At the same time, the Bloom filter matrix adds a marker bit behind each Bloom filter to support arbitrary attribute combination query.

2.3   Cuckoo Hash and Cuckoo Filter

Cuckoo hashing [15] is a simple data structure with the worst case constant lookup time, equaling the theoretical performance of the classic dynamic perfect hashing scheme of Dietzfelbinger et al. [8]. *Cuckoo hashing* is proposed to resolve the hash collision by kicking out the element from its original position to alternate position. It takes up less space and has a faster query speed. It has two hash functions to get possible positions, a hash table and a constant indicating the maximum number of kicks. For each element we need to use these two hash functions to calculate separately to get the two possible positions while inserting, deleting or searching. When insert an element, firstly we have to calculate its two possible positions by using these two hash functions. If both positions are empty, we randomly choose one to insert. If only one position is empty, we chose the empty cell to insert. Otherwise, we randomly choose one position and kick out the original element. The kicked-out element then should be relocate into its another possible location. At the same time, if the number of kicks is greater than 0, we should decrease the number of kicks by one. Otherwise discard the currently kicked-out element. Because each element has two possible locations which are determined by two hash functions, answer a query is just to look up its possible positions and judge whether the position contains the element.

Figure 1 shows the process of inserting x in Cuckoo hash. Using two hash functions, we get its possible positions 1 and 6. But both are occupied. Then we choose position 6 randomly to insert. So x is inserted into position 6, and the original element c has to relocate. And if the number of kicks is greater than 0, decrease it by one. Otherwise discard the currently kicked-out element c. Because its another possible position is 3. So c is relocated into position 3

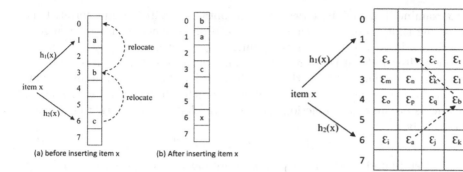

Fig. 1. Cuckoo hash   **Fig. 2.** Insert in Cuckoo filter

and the original element b has to relocate. And if the number of kicks is greater than 0, decrease it by one. Otherwise discard the currently kicked-out element b. Because b's another possible position is 0 which is empty, insert b into position 0.

As shown in Fig. 2, a basic cuckoo filter [9] consists of an array of buckets, with each bucket having multiple entries. What it stores is not the element itself but its fingerprint, a bit string derived from the original item using a hash function. ϵ_x, ϵ_a, ϵ_j in Fig. 2 are fingerprints. Since the fingerprint takes up less space than its initial element and uses hash to insert, the cuckoo filter has the advantages of less space occupation and fast query speed. The main methods of inserting, searching and deleting is similar to cuckoo hash. One of the difference is each position could have many items instead of only one. When insert an fingerprint, first we have to calculate its two possible buckets by using these two hash functions. If both buckets are not full, we randomly choose one bucket and insert the fingerprint to one of its empty entry. If only one bucket is not full, we chose it and insert the fingerprint to one of its empty entry. Otherwise, we randomly choose one bucket and randomly kick out one of the original entries. The kicked-out fingerprint then should be relocated into its another possible bucket. At the same time, if the number of kicks is greater than 0, we should decrease the number of kicks by one, otherwise, discard the currently kicked-out fingerprint. When searching or deleting, we also need to calculate the two possible buckets first, and then traverse in both buckets. To be specific, the positions of the two candidate buckets to store the fingerprint of x are computed by the following equation:

$$h_1(x) = hash(x)$$
$$h_2(x) = h_1(x) \oplus hash(\epsilon_x) \tag{1}$$

Thus, when an element is inserted, searched or deleted, first we have to calculate its fingerprint using some hash function. The insertion process continues until all fingerprints have their own storage location or the relocation times reach a threshold $MaxLoop$. To judge whether an item belongs to the filter, we just need to examine if one of these two buckets contains the item's fingerprint. It is worth noting that there has a chance of false positive when a matching finger-print is found. It means if the corresponding fingerprint is found, the item most likely in the filter and if it is not found, the item must be not in the filter. This is because different item may have the same fingerprint. This is unavoidable, what we can do is just adjusting corresponding parameters to let the false positive rate as low as possible. The deletion removes a random entry of either candidate buckets simply.

$$\epsilon_{CF} = 1 - (1 - 1/2^f)^{2b} \approx 2b/2^f \tag{2}$$

The upper bound of the false positive rate can be computed by Eq. (2). f means the length of each fingerprint is f bits. b is the max number of entries in each bucket and it is the same for each bucket. Because each fingerprint has two possible buckets. So each look up query has to search up to $2 \times b$ positions. Compared with Bloom filter, the Cuckoo filter has apparent advantages. The support of delete operation is the major advantage, and it also has better lookup performance, higher space efficiency when the target false positive rate is less than 3% [9].

3   Multiple Cuckoo Filter

Because bloom filter is not applicable in complex and comprehensive member-ship query, some researches make changes based on the bloom filter to obtain efficiently membership query [19] which create a bloom filter for each dimension. For multiple datasets, the solution is similar. In massive data, it is very difficult to query whether an element exists in all data streams within a certain period of time. Inspired by this, we design a Multiple Cuckoo Filter (MCF), which is based on the classic cuckoo filter to judge if an item is in multiple data streams within a certain period of time.

The MCF consists of a number of standard cuckoo filters. Each data stream corresponds to a cuckoo filter. As such, there are as many filters as the number of data streams. By this way, we can divide the membership query into multi-ple single data stream query. To make this method applicable in multiple data streams, it must be used in conjunction with the window mechanism. Each data stream corresponds to a filter, which stores the fingerprint of all elements in the current sliding window. Firstly we need to split all the data in this period according to the window. Each data stream is split in this way, and the window size of each data stream can be different. Then insert the data in each window into the cuckoo filter of the data stream, the insertion method is the same as the above cuckoo filter. The sliding window uses a queue data structure, because when dynamically changing the sliding window, the head and tail of the window need to be deleted and added separately. The window slides down from the top of the data stream. If the sliding interval is the same as the window size, the sliding window becomes a rolling window.

Figure 3 is a simple schematic of MCF proposed in this paper. In Fig.3, $S1,S2...Sn$ represent different data streams. And here we suppose that within

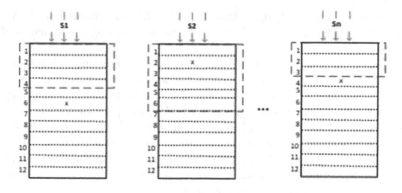

Fig. 3. MCF

a certain period of time, each data stream has exactly 12 records. The dashed box indicates the current processing window of each data stream. The size of the window in the data stream $S1$ is 4, which means we need to process 4 records at once. The size of the window in the data stream $S2$ is 6. The window size in the data stream Sn is 3. Then for each window, we need to insert all data in the window into its own cuckoo filter, the insertion method is the same as cuckoo filter described in last section.

3.1   Checkup Algorithm

The Checkup Algorithm is shown in Algorithm 1. It has two incoming parameters num and x. num indicates the number of data streams and x means we need to judge if the item x can be found in all data streams within a certain period of time. In short, for each data stream, the processing steps are as follows:

Step 1: The sliding window starts from the starting position of the data stream to be processed, and divides the data stream according to the window size and sliding interval;

Step 2: If the current cuckoo filter is empty, insert all records' fingerprint in the window into its own cuckoo filter; If it is not empty, clear the cuckoo filter, and then insert the data in the window into the cuckoo filter;

Step 3: Check whether x exists in the current window. If it does, the current window stops sliding and returns *true*, otherwise go to step 4;

Step 4: Judge whether the current window has been slid to the end of the current data stream. If it is, return *false*. Otherwise the window slides down, the sliding distance is the size of the set sliding interval, and then go to step 3;

Step 5: Judge whether each data stream returns *true*. If all data streams return *true*, it means x does indeed exist in each data stream. But if any data stream returns *false*, it means that the element x does not exist in all data streams within a certain period of time.

Because we use cuckoo filter to store every record's fingerprint in step 2, we just need very little search time to judge if the item x exists in the current window in step 3. And we use the strategy that once found, return immediately. And we only store fingerprint instead of item itself. So we can use less space, and the query speed is fast.

In Algorithm 1, $window[num]$ in line 1 indicates the window deque for each data stream. The sliding window uses a queue data structure, because when dynamically changing the sliding window, the head and tail of the window need to be deleted and added separately. $bottomIndex[num]$ in line 2 stores the bottom index of each data stream within a certain period of time. $windowSize[num]$ in line 2 stores window size of each data stream. $item\_index[num]$ in line 3 indicates the index of x in each data stream. $window\_back[num]$ in line 3 stores the next index of the current window in each data stream. $move\_num[i]$ in line 17 means the sliding interval of current window.

Algorithm 1. Checkup(num, x)

Inputs: num for datastreams, x for checkup.
Outputs: true or false.
1: deque $window[num]$;
2: size_t $bottomIndex[num]$, $windowSize[num]$;
3: size_t $item\_index[num]$, $window\_back[num]$;
4: **for** (each $i \in 0:num$) **do**
5:     $window[i]$.initialize();
6:     $window\_back[i] \leftarrow$ GetNextWindowIndex();
7: top:
8:     $filter[i]$.clear();
9:     $filter[i]$.Add($windowSize[i]$);
10:    $item\_index[i] \leftarrow$ GetItemIndex(x);
11:    **if** $item\_index[i]$!= NULL **then**
12:        **return** true;
13:    **end if**
14:    **if** ($window\_back[i]+1 \geq bottomIndex[i]$) **then**
15:        **return** false;
16:    **end if**
17:    **for** ($j \leftarrow 0...move\_num[i]$ && $window\_back[i]+1 \leq bottomIndex[i]$) **do**
18:        $window[i]$.pop_front();
19:        $window[i]$.push_back($window\_back[i]$++);
20:    **end for**
21:    **goto** top.
22: **end for**

Furthermore, similar to the standard cuckoo filter, MCF proposed in this paper also has false positive rate. And it ensures no false negatives as long as bucket overflow never occurs. The false positive rate is because different elements may produce the same fingerprint after hashing. It is inevitable. But we believe that the low false positive rate can be allowed in massive data.

3.2   False Positive

Without considering the dynamic windows, according to the checkup operations of MCF, each element's query requires the retrieval of all the cuckoo filters. Each data stream has a cuckoo filter. Therefore, the false positive rate of the MCF refers to the probability that at least one cuckoo filter is misjudged. Assuming that the false positive rate of one cuckoo filter is ϵ_{CF}, then in all s cuckoo filters, the probability that there is no misjudgment is $(1 - \epsilon_{CF})^s$. Therefore, the upper limit of the false positive rate for s cuckoo filters combinations is $1 - (1 - \epsilon_{CF})^s$.

At the same time, taking the dynamically changing windows of the cuckoo filter into account. Suppose that each data stream has m records in total, the sliding window size of each data stream is w, and k elements are moved each time. So a total of $\lfloor (m - w + 1)/k \rfloor + 1$ sliding windows is generated. Thus in the dynamically changing windows of s cuckoo filters, actually $s \cdot (\lfloor (m - w + 1)/k \rfloor + 1)$ times is compared, the false positive rate of the MCF should be:

$$\epsilon_{MCF} = 1 - (1 - \epsilon_{CF})^{s \cdot (\lfloor (m-w+1)/k \rfloor + 1)}$$

$$\approx \frac{2bs(m - w + 1)}{k \cdot 2^f}. \tag{3}$$

In Eq. (3), f means the length of each fingerprint is f bits. b is the max number of entries in each bucket and it is the same for each bucket. According to the relationship of the above formula, it can be seen that the false positive rate of the MCF is related to the size of the bucket, the number of cuckoo filters, the total number of collection elements, the size of the sliding windows, the movement size of each window, and the length of the fingerprint. Specifically, longer fingerprint f can significantly reduce the false positive rate ϵ_{CF} and ϵ_{MCF}, but it will also increase space occupancy greatly. In general, the length of fingerprint has an important impact on Cuckoo Filter and MCF.

4   Experiments

We utilize C++ to implement MCF proposed in this paper. For each query, we use *CityHash*[1] to generate a 64-bit random hash value for every element. The high 32 bits are used to locate the bucket index while the low 32 bits to generate the fingerprint. And *MurmurHash* [2] is used to calculate the spare location.

We assume that n data stream sets S $= (x_1, x_2, ... , x_n)$ with no duplicated records, and m represents the total number of records in each set. In order to facilitate the theoretical analysis, all records in the data streams are not changed in the experiments. The records and the size of the sliding window can be set in advance or generated randomly. The s sliding windows move down simultaneously and return *true* if the corresponding fingerprint is found in MCF, otherwise return *false*.

The total number of elements in the experiment is initially set to 1,000,000. The sliding window size should not be set too large as the larger the sliding window is, the more fingerprints are included at any given moment, which may cause unnecessary overhead. As such, the size of each sliding window is set as 2,000. If one of the sliding windows has reached the maximum value and the item to be queried is not found in this window, MCF returns *false* immediately. Because once the item to be queried is not found in some sliding window, it must not be in all data streams within the current time period.

4.1   Checkup

Increase the number of data streams from 1 to 4, that is, the total number of cuckoo filters increases from 1 to 4, and the program randomly generates the specified element for query. Every table concludes 3 groups and every group has 100 queries. Because every element has two candidate buckets, even the same elements, their corresponding and storage location may be different.

Table 1. One cuckoo filter with 100 queries per group.

Matrics	1	2	3
Min time (ms)	0	0	0
Max time (ms)	335	345	298
Average time (ms)	98.5	97.78	85.28
Success count	100	100	100

As shown in Table 1, there is only one data stream. Every data stream have a cuckoo filter, so there is only one cuckoo filter. Here the element to be queried is set to a random generated data, so every query will return *true*, i.e the success rate is 100%. The main purpose of this set of experiments is to prove that the MCF proposed in this paper can search for the item successfully as long as it exists.

When the number of cuckoo filters is 2, there will still be instantaneous query, but the number of successful queries is only about 1/3 of the total, and the average time is nearly 4 times that of one cuckoo filter as shown in Table 2.

Table 2. Two cuckoo filters with 100 queries per group.

Matrics	1	2	3
Min time (ms)	0	0	0
Max time (ms)	1023	991	874
Average time (ms)	400.88	368.33	336.58
Success count	34	46	36

The number of cuckoo filters increases again. From Table 3 we can see it is difficult to find the elements instantaneously. The successful checkup count is getting smaller and smaller, the average time is getting bigger and bigger, and it is nearly 8 times that of one cuckoo filter. It can be seen that the checkup time of 1 to 3 cuckoo filters increases exponentially with a factor of 2.

Table 3. Three cuckoo filters with 100 queries per group.

Matrics	1	2	3
Min time (ms)	521	472	274
Max time (ms)	1525	1589	1271
Average time (ms)	898	994.57	855
Success count	6	7	10

As shown in Table 4, when the number of cuckoo filters is 4, it can be found that only about 3 times per 100 queries are successful, and the success rate drops to 3%.

Table 4. Four cuckoo filters with 100 queries per group.

Matrics	1	2	3
Min time (ms)	520	1	38
Max time (ms)	1545	1457	1632
Average time (ms)	1088.25	821.33	1056.33
Success count	4	3	3

As can be seen from the data of the above figures, when the number of cuckoo filters increases from 1 to 4, the number of successful queries is greatly reduced, and the check-up time increase linearly. When the number of cuckoo filters become more and more, it is difficult to successfully checkup the fingerprint of the specified element at any time within the sliding windows. The more random hash functions are, the less likely of a hash collision happens, and the smaller false positive rate the MCF has.

4.2 Insert and Contain

As shown in Fig. 4, the Insert operation adds elements into the cuckoo filter, and the Contain operation retrieves whether some certain elements are included in the filter. Increase the total number of cuckoo filters from 1 to 10, it can be seen that as the number of cuckoo filters increases, the insertion and query time increase linearly. The growth rate of the insert operation is higher than 40% of the query operation, because kicking and relocating may occur during insertion.

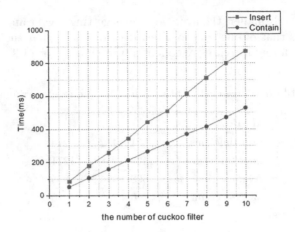

Fig. 4. Insert and Contain time and the number of cuckoo filters.

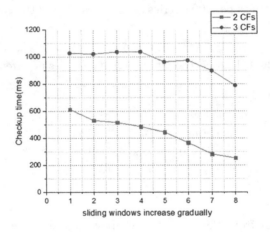

Fig. 5. Checkup time and the number of sliding windows.

4.3   Windows

When examining the relationship between Checkup time and the size of sliding windows, the size of the sliding windows is no longer randomly generated. The maximum size of sliding window equals to the number of total elements. The smaller window size is, the more the number of windows is. The size of sliding window decreases gradually from maximum, so the number of sliding windows is bigger and bigger. From the experimental results, as shown in Fig. 5, it can be seen obviously that the Checkup time decreases as the number of sliding windows increases when there are 2 cuckoo filters and 3 cuckoo filters. When the number of cuckoo filters is larger, the check up time of some element is also significantly increased. In a word, between the same number of cuckoo filters,

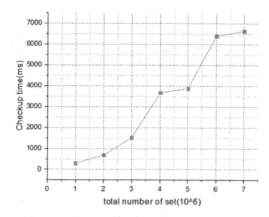

Fig. 6. Checkup time and the sizes of total number.

the smaller the sliding window is, the bigger the number of sliding windows is and the smaller the Checkup time is.

4.4   Total Number of Sets

Since the type of $size\_t$ can represent a range from 0 to 4,294,967,295, the number of set elements is increased from 10^6 to 7×10^6, each time incremented by 10^6. As the total number of elements increases, the number of sliding windows that changing with time also increases. And the number of Checkup comparisons increases, it will result checkup time to increase. When the total number of elements is getting bigger, if the sliding window doesn't increase, the chance of successful query in each sliding window will decrease. So the Checkup time will obviously increase with the size of total number increasing. As shown in Fig. 6.

5   Conclusion

This paper proposes MCF, a multiple cuckoo filter based on the classic Cuckoo filter. The method decomposes the representation and membership query of multiple datasets into multiple single operations, and puts the query in a streaming environment. How much data streams are generated determines how many standard cuckoo filters are used. Our experimental results show that, as the number of data streams increases, the time for insertion and contain operation also increases. When the number of cuckoo filters i.e. data streams number is between 1 and 3, the checkup time for the specified elements presents a clear upward trend. The checkup time consumption of the MCF also increases gradually with the decrease of the sliding window size and the growth of total number. Compared with the traditional cuckoo filter, the MCF can judge if an element exist in all data streams within a certain period of time. It will have a promising and wide range of applications in future.

Acknowledgment. The authors would like to thank the anonymous reviewers for their valuable comments and helpful suggestions, which greatly improved the quality of this paper. This work is supported in part by Key-Area Research and Development Program of Guangdong Province (2020B010164002), Shenzhen Strategic Emerging Industry Development Funds (JCYJ20170818163026031), and also in part by National Natural Science Foundation of China (61672513), Science and Technology Planning Project of Guangdong Province (No. 2019B010137002), and Shenzhen Basic Research Program (JCYJ20170818153016513).

References

1. https://www.cityhash.org.uk
2. Appleby, A.: Murmurhash (2008). https://sites.google.com/site/murmurhash
3. Bender, M.A., et al.: Don't thrash: how to cache your hash on flash. VLDB Endow. **5**, 1627–1637 (2012)
4. Bloom, B.H.: Space/time trade-offs in hash coding with allowable errors, vol. 13, pp. 422–426. ACM (1970)
5. Bonomi, F., Mitzenmacher, M., Panigrahy, R., Singh, S., Varghese, G.: An improved construction for counting bloom filters. In: Azar, Y., Erlebach, T. (eds.) ESA 2006. LNCS, vol. 4168, pp. 684–695. Springer, Heidelberg (2006). https://doi.org/10.1007/11841036_61
6. Breitinger, F., Baier, H.: Similarity preserving hashing: eligible properties and a new algorithm. In: Rogers, M., Seigfried-Spellar, K.C. (eds.) ICDF2C 2012. LNICST, vol. 114, pp. 167–182. Springer, Heidelberg (2013). https://doi.org/10.1007/978-3-642-39891-9_11
7. Debnath, B., Sengupta, S., Li, J.: FlashStore: high throughput persistent key-value store. Proc. VLDB Endow. **3**(1–2), 1414–1425 (2010)
8. Dietzfelbinger, M., Karlin, A.R., Mehlhorn, K., Heide, F.M.A.D., Tarjan, R.E.: Dynamic perfect hashing: upper and lower bounds. In: Symposium on Foundations of Computer Science (1988)
9. Fan, B., Andersen, D.G., Kaminsky, M., Mitzenmacher, M.D.: Cuckoo filter: practically better than bloom. In: Proceedings of the 10th ACM International on Conference on emerging Networking Experiments and Technologies, pp. 75–88. ACM (2014)
10. Fan, L., Cao, P., Almeida, J., Broder, A.Z.: Summary cache: a scalable wide-area web cache sharing protocol, vol. 8, pp. 281–293. IEEE (2000)
11. Guo, D., Wu, J., Chen, H., Luo, X.: Theory and network applications of dynamic bloom filters, pp. 1–12 (2006)
12. Kornblum, J.: Identifying almost identical files using context triggered piecewise hashing. Digit. Invest. **3**, 91–97 (2006)
13. Lynch, C.: Big data: how do your data grow? Nature **455**, 28–9 (2008)
14. Nishimura, S., Das, S., Agrawal, D., El Abbadi, A.: MD-HBase: a scalable multidimensional data infrastructure for location aware services, vol. 1, pp. 7–16 (2011)
15. Pagh, R., Rodler, F.F.: Cuckoo hashing, pp. 121–133 (2001)
16. Putze, F., Sanders, P., Singler, J.: Cache-, hash- and space-efficient bloom filters. In: Demetrescu, C. (ed.) WEA 2007. LNCS, vol. 4525, pp. 108–121. Springer, Heidelberg (2007). https://doi.org/10.1007/978-3-540-72845-0_9
17. Rodeh, O., Teperman, A.: zFS-a scalable distributed file system using object disks. In: 2003 Proceedings of the 20th IEEE/11th NASA Goddard Conference on Mass Storage Systems and Technologies (MSST 2003), pp. 207–218. IEEE (2003)

18. Roussev, V.: Data fingerprinting with similarity digests. In: Chow, K.-P., Shenoi, S. (eds.) DigitalForensics 2010. IAICT, vol. 337, pp. 207–226. Springer, Heidelberg (2010). https://doi.org/10.1007/978-3-642-15506-2_15
19. Wang, Z., Luo, T., Xu, G., Wang, X.: A new indexing technique for supporting by-attribute membership query of multidimensional data. In: Gao, Y., et al. (eds.) WAIM 2013. LNCS, vol. 7901, pp. 266–277. Springer, Heidelberg (2013). https://doi.org/10.1007/978-3-642-39527-7_27
20. Wang, Z., Luo, T., Xu, G., Wang, X.: The application of cartesian-join of bloom filters to supporting membership query of multidimensional data, pp. 288–295 (2014)
21. Xie, K., Qin, Z., Wen, J.G., Zhang, D.F., Xie, G.G.: Combine multi-dimension bloom filter for membership queries. J.-China Instit. Commun. **29**(1), 56 (2008)
22. Zhou, X., Zhang, X., Wang, Y., Li, R., Wang, S.: Efficient distributed multi-dimensional index for big data management. In: Wang, J., Xiong, H., Ishikawa, Y., Xu, J., Zhou, J. (eds.) WAIM 2013. LNCS, vol. 7923, pp. 130–141. Springer, Heidelberg (2013). https://doi.org/10.1007/978-3-642-38562-9_14

A Data-Driven Method for Dynamic OD Passenger Flow Matrix Estimation in Urban Metro Systems

Jiexia Ye[1,2], JuanJuan Zhao[1(✉)], Liutao Zhang[1,4], ChengZhong Xu[3], Jun Zhang[1], and Kejiang Ye[1]

[1] Shenzhen Institutes of Advanced Technology, Chinese Academy of Sciences, Beijing, China
{jx.ye,jj.zhao,lt.zhang,jun.zhang1,kj.ye}@siat.ac.cn
[2] Shenzhen College of Advanced Technology, University of Chinese Academy of Sciences, Beijing, China
[3] State Key Lab of Iotsc, Department of Computer Science, University of Macau, Macau, China
czxu@um.edu.mo
[4] University of Science and Technology of China, Hefei, China

Abstract. Dynamic O-D flow estimation is the basis of metro network operation, such as transit resource allocation, emergency coordination, strategy formulation in urban rail system. It aims to estimate the destination distribution of current inflow of each origin station. However, it is a challenging task due to its limitation of available data and multiple affecting factors. In this paper, we propose a practical method to estimate dynamic OD passenger flows based on long-term AFC data and weather data. We first extract the travel patterns of each individual passenger based on AFC data. Then the passengers of current inflows based on these patterns are classified into fixed passengers and stochastic passengers by judging whether the destination can be inferred. Finally, we design a K Nearest Neighbors (KNN) and Gaussian Process Regression (GPR) combined hybrid approach to dynamically predict stochastic passengers' destination distribution based on the observation that the distribution has obvious periodicity and randomicity. We validate our method based on extensive experiments, using AFC data and weather data in Shenzhen, China over two years. The evaluation results show that our approach with 85% accuracy surpasses the results of baseline methods and the estimation precision reaches 85%.

Keywords: Metro systems · AFC data · Data mining · Intelligent transportation systems · Dynamic OD flow

1 Introduction

Dynamic OD flows reflect the traffic demand in a metro system. They describe the number of passengers of all O-D (origin station - destination station) pairs of the metro system, and they are the basis of traffic control and management, such as capacity allocation, emergency coordination. OD traffic flows have time delay problem, which

S. Nepal et al. (Eds.): BIGDATA 2020, LNCS 12402, pp. 116–126, 2020.
https://doi.org/10.1007/978-3-030-59612-5_9

refers that current inflow of a station will leave the system in the future time from other stations. In this paper, we aim to predict the destination distribution of current inflow at each station.

Many works have been done to predict individual user's destination based on spatial sequence patterns, which mine the sequential frequent patterns from the user's historical trajectories [1–4]. Such patterns can reflect co- occurrence of locations. These are the places that the user always visit after visiting somewhere. For example,

Morzy et al. [1] used an Apriori-based algorithm to generate association rules from the moving object database and used it for user's location prediction. Morzy et al. [1] used a modified version of PrefixSpan algorithm to discover frequent movements of users [2]. Zheng et al. [4] used a HITS-based inference model to mine users' interesting location and detected users' travel sequence to make locations prediction. Overall, the destination prediction of the existing approaches relies on transition probabilities between different locations based on historical trajectories using various Markov chain-based models. It is hard to directly apply these methods to our problem as follow: One of the key prerequisites is that there should be enough historical data of individuals to learn the transition probabilities. However, in our context, there are two third passengers only have limited historical data, which makes it hard to directly apply these methods mainly designed for individuals instead of group.

Recently, thanks to the upgrades of urban infrastructures, we can obtain long-term data collected by AFC system, where all passengers need to tap their smart cards when they enter or leave the metro system, which brings us opportunity to understand passenger OD flows from the raw data. By now, AFC data has been used to inform transit agency from planning, to operations, to measuring and monitoring performance [5–8], and understanding user spatiotemporal travel patterns [9–12]. Different from them, in this paper, we focus on a basic problem, dynamic OD flow estimation.

In this paper, we propose a novel method of estimating the dynamic OD flow in metro systems based on long-term AFC data and weather data. Our intuition is that the passengers in metro system can be categorized into two groups, fixed passengers and stochastic passengers. The fixed passengers are spatiotemporal regular, such as commuters, students. Their travel behaviors are relatively stable. They often travel from one fixed station to another during relatively fixed time period. Their destination can be uniquely inferred based on their travel patterns mined from their own long-term smart card records. While stochastic passengers are spatial temporal irregular, based on the observations (detailed in following section) that their destination distribution has obvious regularity (e.g. 24-h periodicity, and difference between weekday and weekend) and randomicity (affected by some factors, e.g., weather, past traffic flows), we propose a KNN and GPR hybrid model to estimate their destination distribution. The major contribution of this paper is threefold:

1. We propose a novel method to estimate the dynamic OD flows in a metro system. The method first uses a kernel density-based clustering algorithm to extract each passengers' travel pattern from long-term AFC data. Then we classify all passengers of current inflow into two groups, fixed passengers and stochastic passengers according to whether their destination can be uniquely inferred based on their travel patterns.

2. For efficiently and robustly predicting the destination distributions of stochastic passengers, we design a dynamic KNN and GPR combined approach with two types of data, the AFC data and weather data, based on the insight that their destination distribution has obvious periodicity and randomicity.
3. We evaluate our method using real-world data of smart card data and weather data over two years in Shenzhen, China. The experimental results show the effectiveness of our method.

The rest of this paper is organized as follows. Section 2 presents the overview of our problem and solution. Section 3 details our solution. The empirical evaluation of the method on real-world data is discussed in Sect. 4, and we conclude the study in Sect. 5.

2 Overview

2.1 Notations and Assumptions

We first present several useful definitions:

Definition 1 (Metro System): A metro system is composed of multiple lines $L = \{l_1, l_2, ..., l_{|L|}\}$ and multiple stations $S = \{s_1, s_2, ..., s_{|S|}\}$, where $|L|$ and $|S|$ is the number of metro lines and metro stations. A metro line l_i passes by some ordered stations.

Definition 2 (Time slots): We divide a day into multiple periods $T = \{T_1, T_2, ..., T_{|T|}\}$ by a fixed interval τ. The k period contains a time range of $\{(k-1)\tau, k\tau\}$.

Definition 3 (Trip): A trip of a passenger, denoted as Tr, is related to four attributes s_o, s_d, t_o, t_d, representing tap-in station, tap-out station, tap-in time and tap-out time. A trip is obtained by joining the tap-in and tap-out records collected by AFC system through matching card ID and time.

Definition 4 (Inflow): We use $O_{i,k}$ to represent the passengers flows entering metro system from station s_i during time period T_k.

Definition 5 (OD flow): The passengers who enter a metro network from station s_i during time interval T_k and destine to station s_j, denoted as $OD_{i,k}^j$.

2.2 Problem Definition

In this paper we use two types of data: smart card transaction data and weather data.

Smart card data: Metro passengers need to tap their smart card when they enter and exit metro system. Each record includes card ID, stations, transaction time and transaction type (tap-in and tap-out).

Weather data: We use weather data collected by district-level weather stations. Each record consists of time of day, rainfall, temperature, humidity, wind speed and wind direction.

Dynamic OD flow estimation: Given real-time smart card data, weather data, and current time slot T_c, this paper aims to estimate current OD flow matrix $M_{|S| \times |S|}$, where the item of m_{ij} is the ODflow $OD_{i,c}^j$, representing the number of the passengers of $O_{i,c}$ who will destine to station s_j.

2.3 Data-Driven Insight

By now, there has been a lot of researches focusing on understanding the travel patterns of individual users in urban cities [12–14]. We conclude from these studies that the trips of passengers can be classified into two types, fixed trips and stochastic trips. The fixed trips of a user refer that the user often travels from one fixed place to another fixed place at fixed time interval of a day. Thus, passengers can be regard as fixed and we can infer the destination if given the origin station. Aside from fixed trips, we find that the destination distribution of stochastic trips has some regularity and randomicity as follows.

First, historical data has high reference for real-time OD flow estimation. We take 30 min as granularity and calculate the correlation coefficient of OD flows between weekdays, between weekends. The result is shown in Fig. 1. We found that there are strong correlations since the average correlation coefficient are greater than 0.5. The finding indicates that historical data has high reference for real-time dynamic OD flow estimation. In addition, we calculate the similarities of the destination distributions by cosine distance, between rainy day and sunny day, between rainy days, and between sunny days at the same time slot of day. They are 0.51, 0.601, 0.634 respectively. It means that the more similar the weather conditions between two days are the more similar the destination distributions are.

Second, the stochastic passengers' destination distribution also has randomicity by these external factors (e.g. weather, real-time traffic flows), which is dynamic and we can only give a reliable prediction by using the most updated model.

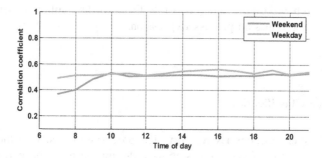

Fig. 1. Correlation coefficient of passenger destination distribution in different days of week

2.4 Analysis Framework

Figure 2 presents the framework for dynamic OD-Matrix predictive method. First, based on the long-term history smart card transaction data, we extract the spatialtemporal travel patterns of individual passengers. Based on these patterns, the passengers of inflows are divided into two categories according to whether their destination can be uniquely inferred: fixed passengers and stochastic passengers. Then we use a KNN and GPR hybrid model to estimate the proportion that the stochastic passengers will destine to other stations. Finally the OD matrix is obtained by combining the destination estimation results of fixed passengers and random passengers.

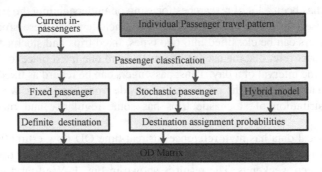

Fig. 2. Processing flowchart

The reason why we use KNN and GPR hybrid model is as follow. The OD flow of stochastic passengers is dynamic and has randomicity, we can only give a reliable prediction by using the most updated model for prediction. This implies that frequent retraining is necessary. To fulfill this requirement, we propose a lazy learning approach that integrates the KNN and GPR (Gaussian process regression) for efficient and robust OD flow prediction. GPR is known to be efficient in dealing with relationships among data. However it cannot be scaled up easily for its cubic learning computation and quadratic space requirement. Given the Gaussian process computation barrier, we have to solve a small-scale problem by selecting the limited influential data for GPR, so KNN is combined based on the regularity of destinations distribution, to select the most informative data for Gaussian process Regression.

3   Solution

3.1   Passenger Classification

Individual's Travel Patterns Extraction. The process of classifying the passengers who have entered a metro system at station s_i during current time period I_c can be divided into two stages: individual passenger's travel patterns extraction, dynamic passenger classification.

Definition (Travel pattern): a travel pattern p of individual user is used to describe the travel's spatio-temporal regularity, which refers that the passenger often goes from one fixed station to another fixed destination station during a fixed periods of days, and the proportion of these trips exceeds a specified percentage threshold λ (e.g., 60%). For example, in 60% of the days, Bruce goes from the University Town to Shenzhen North Station between 8: 30 and 9: 30. A travel pattern $P = \{s_o, s_d, t_o, t_d, r\}$ contains the origin station s_o, destination station s_d, start time t_o, end time t_d and proportion r respectively.

Given all trips $TR = \{tr_1, tr_2, \ldots, tr_{|N|}\}$ of a passenger, we use two steps to extract the passenger's travel patterns. First, we use a kernel density-based clustering algorithm to cluster passengers' history trips according the time points of a day. Then, for each cluster, we find whether there exists a subset of trips with similar OD pair, and the proportion of such trips to the total days is greater than the threshold λ. If such subset does not exist, no travel pattern is constructed; Otherwise a travel pattern is constructed. In the following, we mainly focus on the first step.

Due to the fact that different passengers have different number of travel patterns, for example, some passengers have one fixed travel period during a day, some passengers have two or more fixed travel periods during a day. Therefore, the number of clusters can't be determined at the beginning of clustering. Here we use the kernel density-based cluster algorithm which can automatically determine the number of clusters.

The basic principle of the kernel density-based cluster algorithm is that an ideal cluster center has two characteristics: 1) compared with the neighboring data points, the center point of a cluster has greater local density ρ; 2) compared with other data points, the distance σ between the centers of clusters is relatively large. This algorithm is not only suitable for cluster analysis of large-scale data, but also can quickly reduce noise points.

Given a trip set TR, we use the middle time $t_m = \frac{1}{2} \times (tr.t_o + tr.t_d)$ of a trip tr to represent the trip's time for clustering. We assume the time point x of a trip is an independently distributed random variable, the kernel density estimation function is as follows.

$$f(x; h) = \frac{1}{nh} \sum_{i=1}^{n} K\left(\frac{x - x_i}{h}\right) \tag{1}$$

where $K\left(\frac{x-x_i}{h}\right)$ is the kernel function; h is the bandwidth, which determines the size of the local range affected by the kernel function.

For the travels in each cluster, if there is a travel subset that satisfies the conditions mentioned earlier, the starting station and destination station of this subset are considered as the of the travel pattern and the time periods involved in these travels are considered as the of the travel pattern. Based on this method, we extract the travel pattern of each passenger.

Through the above process, we can extract the travel pattern of each passenger, denoted as $P = \{p_1, p_2, \cdots, p_{|P|}\}$.

Online Passenger Classification. Given a passenger of $O_{i,c}$ of a station s_i, who enters metro system in a station s_i at current time slot T_c, we first check the passenger's travel pattern M. If there exists a travel pattern m satisfying $[m.t_o\ m.t_d] \cap T_k \neq null$, we set $m_i.d$ as the destination for the passenger, and classify the passenger as a fixed passenger, otherwise the passenger is a stochastic passenger.

3.2 Destination Distribution Estimation for Stochastic Passengers

For stochastic passenger, we use a hybrid model of *KNN* and *GPR* to model the linear and nonlinear characteristics based on different input features respectively. For the stochastic passengers who enter metro system at station s_i during current time period T_c, we predict the proportion of these stochastic passengers which will destine to other stations, denoted as a vector $Y_{i,c} = \left\{ Y_{i,c}^1, Y_{i,c}^2, \ldots, Y_{i,c}^{|S|} \right\}$.

Input of Model. Given a station s_i and a time slot T_c of a day, we explore three sets of key features, namely time features, weather features, passenger flow features for predicting the proportion of the stochastic passengers travelling to other stations.

Time Features: The time of day and day of the week are significant in OD flow prediction as shown in Fig. 1. We use $F_t = (dy, Ts)$ to represent the time features, where $dy \in [1, 7]$ represents the day of week, which distinguishes between weekdays from Monday to Sunday. The $Ts \in [1, |T|]$ represents the time slot of day.

Weather Features: We use two weather features, tm (temperature) and hm (humidity), which are studied being correlated with the estimated target. Given current time slot T_c needed to estimate, we use the weather condition $F_w = (tm_c, hm_c)$ to represent the weather features.

Traffic Flow Features: We extract the inflow and outflow of each metro station during past time M intervals, which refers to the total number of passengers who enter and exit metro system at each station during past time interval, $F_l = \{F_{c-M}, \ldots, F_{c-1}\}$, where $F_{c-i} = \{O_{c-i}, D_{c-i}\}$ represents the traffic flow at past time interval T_{c-i}.

K-nearest Neighbor. The KNN is used to find the data close to current condition. An efficient prediction is based on how we can catch the spatial temporal patterns from the historical data.

The condition is represented as a feature vector x which includes the three types of features: time, weather, traffic flow, $x = \{F_t, F_w, F_l\}$. The target value Y is probabilities of the stochastic passengers destining to other stations.

Since destination distribution has obvious temporal characteristics, such as 24 h periodicity, different characteristics between weekday and weekend, and strong correlation between near time slots. We only search on the historical data with same day indicator (weekday and weekend) and near time slots.

After that, we can define the metric for KNN by (1) for each pair of historical data F and test data F'. In the following equation, $C(f, f')$ is the correlation coefficients of OD flows during the time features of F and F'.

$$ dist(F, F') = \prod_{f \in F_t} C(f, f') \times \left(\sum_{f \in F_w} |f - f'| + \sum_{f' \in !F_f} |f - f'| \right) \qquad (3) $$

Gaussian Process Regression. The K neighbors selected by KNN algorithm are used as the input of GPR computation for passengers' destination distribution. Given the K nearest neighbors $H \equiv \{X = (x_i, \cdots, x_k), Y = (y_i, \cdots, y_k)\}$, where x_i denotes an input feature vector and y_i denotes the target value. The goal of GPR is to find a function f which can describe the relationship between x_i and y_i. GPR can be completely specified by a mean function $m(x)$ and a covariance function $k(x,x')$, as $f = GP(m(x), k(x,x'))$, where x and x' represent two input feature vectors. The key point of GPR is to select an appropriate kernel function k. In this paper, we take the most commonly used method that is the squared exponential kernel function as follows:

$$k(x,x') = \sigma^2 \exp\left(-\frac{(x-x')^2}{2\ell^2}\right) \tag{3}$$

Given current conditions x_c, the corresponding target is predicted with GPR as:

$$y_c|x_c, X, Y \sim GP\left(k(x_c, X)\left[k(X,X)\right]^{-1}Y, k(x_c, x_c) - k(x_c, X)\left[k(X,X)\right]^{-1}k(X, x_c)\right) \tag{4}$$

4 Analysis of Experimental Results

4.1 Dataset

In the experiments, we use two types of data: (a) AFC data; (b) weather data over two years from Jan, 1, 2014 to Dec, 31, 2015 in Shenzhen, China. The metro system has 117 stations and 5 physical lines.

4.2 Metrics

In this paper, we evaluate the performances of our model by two kinds of evaluation metrics, which are the mean accuracy ACC, mean absolute error MAE, according to following two Equations. Their definitions are as follows, where y is the ground truth of OD flow matrix, $\tilde{y}$ is the prediction of OD flow matrix. To be noted, the absolute value of the subtraction between two matrix is defined as the sum of the absolute differences of all items in the two matrixes.

$$ACC = 1 - \frac{1}{n}\sum_{i=1}^{n}\frac{|\tilde{y}i - yi|}{yi} \tag{5}$$

$$MAE = \frac{1}{n}\sum_{i=1}^{n}|\tilde{y} - yi| \tag{6}$$

4.3 Compared Methods

We evaluate our method by comparing with the other learning models. The selected learning models are presented as follows, and each of these models is a representative of a group of method with similar base:

Empirical Estimation (EMP): represents the method based on the naive empirical knowledge, we estimate dynamic OD flow matrix by using its historical average.

Bayesian Network (Bayes): is a typical graph-based algorithm, which is a representative for the probability based models.

Artificial Neural Network (ANN): represents the model to capture the non-linearity between three type of features and OD flows.

XGBoost: represents the prediction by resemble learning method for regression problems, by producing a prediction model in the form of an ensemble of basic learning models, typically decision trees.

KGmetNC: In order to justify the necessity of the step of passenger classification in our method. We use a method called as KGmetNC that doesn't perform such step. KGmetNC considers all passengers of current inflow as stochastic passengers and uses KNN-GPR combined method to estimate passengers' destination distribution.

4.4   Experiment Result

Comparision to Baselines. We compare our method with various methods in terms of metrics and highlight the best performance with bold font (as shown in Table 1). First, we can see from Table 1 that our method with more than 0.85 accuracy has the best performance. Moreover, the naive empirical baseline achieves more than 0.652 accuracy, which suggests the OD flows are relatively stable. The Ensemble method XGBoost achieves the best performances among all tradition methods including EMP, Bayes, ANN, which suggests that ensemble methods are also a good choice since they can capture the non-linear correlation between dynamic OD flow and observation. Second, we also can observe the contribution of using the step of passenger classification.

Table 1. Comparision with baselines

Method	ACC	MAE
EMP	0.652	8445.454
Bayes	0.667	8133.047
ANN	0.712	7013.824
XGBoost	0.779	5329.671
KGmetNC	0.702	7278.429
KGmet	0.853	3567.994

4.5   Different Time Period

We compare the predicting results during different time periods of days as shown in Table 2. Specifically, we choose two kinds of days, weekend and weekday, and partition a day into two parts, peeking hours (07:00–09:00, 17:00–19:00) and low hours (other times) for comparison. We can observe that the predicting results at low hours on weekdays are worse than those at peeking hours. It is easy to understand for the reason that there are larger amount of fixed passengers at peak hours, especially at AM peak hours. Same reasons for that the average accuracy in weekend is worse than that in weekday.

Table 2. Relatvie mean error of KGmet

Time period	ACC	MAE
Weekday-peek	0.867	4465.412
Weekday-low	0.845	2347.623
Weekend-peek	0.843	3900.915
Weekend-low	0.841	1481.423

4.6   Different Data Sizes

We evaluate the performance of the proposed method using different data sizes. Basically, we want to confirm that more data can really improve the predicting performance. We test the proposed method with different sizes of the historical data, from one month to nearly two years data. Figure 3 shows the RME with data size. In Fig. 3, we can see that more training data indeed gives better performance. However, the result is going to be stable as the number of input instances goes larger than 18 months.

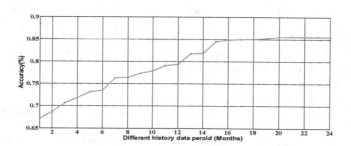

Fig. 3. Using different data size

5   Conclusion and Discussion

In this paper, we propose KGmet, a data-driven method to predict the dynamic OD flow in a complex metro system. KGmet first divides the passengers of current inflow into two categories: fixed passengers and random passengers according to individual passengers' travel patterns. Then KGmet uses a KNN-GPR combined method, which can keep updating based on newly acquired data, to predict the dynamic destination distribution of stochastic passengers. We evaluate KGmet with real-world datasets collected from Shenzhen, China. The experimental results show that our method works well for estimating dynamic OD flows with 85% accuracy.

Acknowledgment. The authors would like to thank anonymous reviewers for their valuable comments. This work is supported in part by the National Key R\&D Program of China (No. 2019YFB2102100), and by "National Natural Science Foundation of China" No. 61802387, and by National Natural Science Foundation of Shenzhen No. JCYJ20190812153212464, and by Shenzhen Discipline Construction Project for Urban Computing and Data Intelligence,and by China's Post-doctoral Science Fund No. 2019M663183.

References

1. Morzy, M.: Prediction of moving object location based on frequent trajectories. In: Levi, A., Savaş, E., Yenigün, H., Balcısoy, S., Saygın, Y. (eds.) ISCIS 2006. LNCS, vol. 4263, pp. 583–592. Springer, Heidelberg (2006). https://doi.org/10.1007/11902140_62
2. Morzy, M.: Mining frequent trajectories of moving objects for location prediction. In: Perner, P. (ed.) MLDM 2007. LNCS (LNAI), vol. 4571, pp. 667–680. Springer, Heidelberg (2007). https://doi.org/10.1007/978-3-540-73499-4_50
3. Jeung, H., Liu, Q., Shen, H.T., Zhou, X.: A hybrid prediction model for moving objects. In: A Hybrid Prediction Model for Moving Objects, pp. 70–79. IEEE (2008)
4. Zheng, Y., Zhang, L., Xie, X., Ma, W.-Y.: Mining interesting locations and travel sequences from GPS trajectories. In: Mining Interesting Locations and Travel Sequences from GPS Trajectories, pp. 791–800 (2009)
5. Zhang, D., Zhao, J., Zhang, F., Jiang, R., He, T., Papanikolopoulos, N.: Last-mile transit service with urban infrastructure data. ACM Trans. Cyber-Phys. Syst. **1**(2), 1–26 (2016)
6. Cheon, S.H., Lee, C., Shin, S.: Data-driven stochastic transit assignment modeling using an automatic fare collection system. Transp. Res. Part C: Emerg. Technol. **98**, 239–254 (2019)
7. Allahviranloo, M., Chow, J.Y.: A fractionally owned autonomous vehicle fleet sizing problem with time slot demand substitution effects. Transp. Res. Part C: Emerg. Technol. **98**, 37–53 (2019)
8. Zhang, D., Zhao, J., Zhang, F., He, T.: coMobile: real-time human mobility modeling at urban scale using multi-view learning, pp. 1–10 (2015)
9. Antoniou, C., Dimitriou, L., Pereira, F.: Mobility patterns, big data and transport analytics: tools and applications for modeling (2018)
10. Zhang, D., Zhao, J., Zhang, F., He, T., Lee, H., Son, S.H.: Heterogeneous model integration for multi-source urban infrastructure data. ACM Trans. Cyber-Phys. Syst. **1**(1), 1–26 (2016)
11. Zhao, J., et al.: Estimation of passenger route choice pattern using smart card data for complex metro systems. IEEE Trans. Intell. Transp. Syst. **18**(4), 790–801 (2016)
12. Zhao, J., Qu, Q., Zhang, F., Xu, C., Liu, S.: Spatio-temporal analysis of passenger travel patterns in massive smart card data. IEEE Trans. Intell. Transp. Syst. **18**(11), 3135–3146 (2017)
13. Kusakabe, T., Asakura, Y.: Behavioural data mining of transit smart card data: a data fusion approach. Transp. Res. Part C: Emerg. Technol. **46**, 179–191 (2014)
14. Widhalm, P., Yang, Y., Ulm, M., Athavale, S., González, M.C.: Discovering urban activity patterns in cell phone data. Transportation **42**(4), 597–623 (2015). https://doi.org/10.1007/s11116-015-9598-x

Ensemble Learning for Heterogeneous Missing Data Imputation

Andre Luis Costa Carvalho$^{(\boxtimes)}$, Darine Ameyed$^{(\boxtimes)}$, and Mohamed Cheriet$^{(\boxtimes)}$

System Engineering Department, University of Quebec's Ecole de Technologie
Superieure, Montreal, QC H3C 1K3, Canada
{andre-luis.costa-carvalho.1,darine.ameyed.1}@ens.etsmtl.ca,
mohamed.cheriet@etsmtl.ca

Abstract. Missing values can significantly affect the result of analyses and decision making in any field. Two major approaches deal with this issue: statistical and model-based methods. While the former brings bias to the analyses, the latter is usually designed for limited and specific use cases. To overcome the limitations of the two methods, we present a stacked ensemble framework based on the integration of the adaptive random forest algorithm, the Jaccard index, and Bayesian probability. Considering the challenge that the heterogeneous and distributed data from multiple sources represents, we build a model in our use case, that supports different data types: continuous, discrete, categorical, and binary. The proposed model tackles missing data in a broad and comprehensive context of massive data sources and data formats. We evaluated our proposed framework extensively on five different datasets that contained labelled and unlabelled data. The experiments showed that our framework produces encouraging and competitive results when compared to statistical and model-based methods. Since the framework works for various datasets, it overcomes the model-based limitations that were found in the literature review.

Keywords: Ensemble methods · Missing data imputation ·
Distributed data · Multidomains · Big data · Smart city

1 Introduction

Missing data is a widespread problem that can affect the quality of any study and produce biased estimates, leading to invalid conclusions. Health, all sorts of businesses, surveys of any nature, and automated sensors are just a few of countless examples [1,2]. Indeed, besides missing values, any system (e.g., application, and platform) is subject to producing data that might be imprecise, insufficient, duplicated, incorrect, and inconsistent [3]. Particularly in analytic and learning processes, such inconsistency can significantly impact the conclusions that are drawn from the data and may bias the decision process. In the literature, one of the possible reasons for the loss of data is missing data, and adequate treatment

© Springer Nature Switzerland AG 2020
S. Nepal et al. (Eds.): BIGDATA 2020, LNCS 12402, pp. 127–143, 2020.
https://doi.org/10.1007/978-3-030-59612-5_10

is usually in accordance with three of the most prominent techniques. 1) Missing at random (MAR): the probability that an attribute will have missing values depends on another attribute or observation. 2) Missing completely at random (MCAR): all attributes will have the same probability of missing values independent of the observation. 3) Missing not at random (MNAR): the missing value in an attribute is strictly dependent on the values in other attributes or observations [4]. Over the past years, the missing values problem has been tackled mostly from either a statistical or model-based perspective.

Statistical Perspective: Solutions from statistical methods, such as the mean, median, arbitrary values, end of the distribution, and most frequent value (MFV) or mode can be straightforwardly applied, the drawback of such methods is the introduction of some level of bias. Furthermore, many of these methods only work with continuous attributes.

Model-Based Perspective: Model-based methods are customized machine learning (ML) algorithms that typically map a proposed solution onto a specific or particular problem. An advantage of this approach is its rapid prototyping to meet a specific demand. The disadvantage, however, is the need to formulate a solution for each new problem [21]. The model-based approach allows for univariate and multivariate analyses. In a univariate analysis, the values in an attribute are used as a reference to impute its missing ones. In multivariate analysis, one attribute might be useful for inputting the missing values in another one, mainly in the case of a high degree correlation [17]. However, a problem of relying on such correlation is that even though two attributes can have a strong correlation, they can represent different information in a dataset.

Missing Values, Further Context: Another critical example of missing values occurs in scenarios of distributed and heterogeneous data (Fig. 1), which is generated by different sources, providers, and formats. It is usually composed of unstructured, semi-structured (e.g., images, streaming, and text files), and structured data (e.g., a relational database) [5]. In addition, the current sophisticated platforms generate a massive amount of heterogeneous data. Therefore, this paper addresses missing values of data from multiple sources, formats and types: images, sensors, synthetic, real-world, labelled, and unlabelled datasets. To do this, considering the variety of data types mentioned above, we propose an ensemble ML framework. Ensemble methods train multiple learners and then combine them. It is a state-of-the-art learning approach. It is well known that an ensemble is usually significantly more accurate than a single learner, and ensemble methods have already achieved great success in many real-world tasks [22].

The remainder of this paper will proceed as follows: We first introduce the problem, motivation, and contributions in Sect. 2. Then, in Sect. 3, we present the notation, related work, and the methodology of our approach. After that, in Sect. 4, we describe our proposed framework. Section 5 presents the performance evaluation. In Sect. 6, we present the performance evaluation and discuss the results. Finally, in Sect. 7, we state the conclusion and make remarks about possible future work.

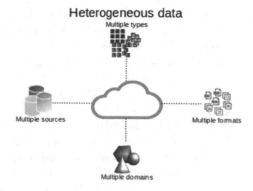

Fig. 1. Distributed and heterogeneous data.

2   Problem and Motivation

2.1   The Problem of Missing Values

A common problem in open and multilple source datasets, data that are generated by different systems, databases, and platforms [8], is that although the sources can be verified, they cannot ensure the quality of the data. Also, various problems like insufficient, duplicated, incorrect, uncertain, ambiguous, and mainly missing values are frequently present [3,8]. Also, in the big data and distributed data context, the datasets are usually composed of different types of data, including continuous, discrete, categorical, and binary. When applying statistical methods to overcome the missing values, other problems can appear. For instance, one essential aspect of an efficient ML model is the level of independence and the identical distribution of the data (i.i.d). Another problem is the use of mean or median to impute missing values in the train data. It can directly affect its distribution, which brings unsatisfactory and unreliable results into the prediction tasks [12,13]. In model-based methods, the approaches are determined by the data type, and they do not cover data heterogeneity. Usually, they are limited in restricted or specific use cases. Thus, the missing data problem remains open and requires even more attention in big data scenario, which involve multiple sources, modalities, and heterogeneous data.

2.2   Motivation to Propose the Ensemble Learning Model

In the context described, our work aims to address the problem of missing values to enhance the power of the analysis and its quality results, keeping the statistical strength of the dataset, which might have this capacity degraded due to missing values. We propose an innovative stacked ensemble framework to reconstruct datasets that suffer from missing values, applying an ML approach instead of the mature techniques already described and addressed in the data quality fields. In the relational database management system (RDBMS) and NoSQL non-relation database, the tables have basic functionalities that reduce anomalies and inconsistencies, protecting data integrity. Specifically, in the RDBMS, through the relationship in the tables or other database schemas, it can be useful in some instances to tackle the missing data problem [23]. On the other hand, in the distributed/isolated datasets, (commonly CSV, XML, JSON files), this is not possible without an external relationship. Therefore, we propose an ML approach to deal with the problem in those distributed/isolated datasets in which the missing values cannot be filled by the relationship, and the algorithm has to learn exclusively from the data that is available in the dataset itself.

We introduce a framework that is composed by combining the random forest (RF) ensemble, which is one of the most widespread and robust ML algorithms with the Jaccard index and Bayes probability to impute missing values based on data topology [1, 20]. Our goal is to provide a robust framework that can preprocess heterogeneous and multiple sources of data to ensure better data quality for data fusion and knowledge extraction, which will transform it into useful information and, subsequently, into knowledge [6, 7]. Taking into account the heterogeneity and multiple sources of data, it necessary to pay close attention to controlling their validity and reliability to achieve this [3]. Therefore, considering that datasets are subject to all sorts of impurities, missing values are still a considerable challenge.

2.3   Contributions of Our Ensemble Framework

Our stacked ensemble framework is a combination of an RF algorithm (which has been significantly adapted for our purposes), Jaccard index, and Bayes probability. Combining these techniques, we produced a robust framework for imputing missing values. The framework's contributions are summarized as follows:

- It increases the quality of the datasets affected by missing values. This is a basic problem faced daily while preprocessing the data in the ML pipeline.
- The proposed framework saves time and effort in data cleaning and data wrangler, which is often a time-consuming step in the ML and data mining (DM) process.
- It decreases the level of lost information in the cleaning phase, producing better results in the further steps of data manipulation for the ML and DM processes.
- Finally, it creates a reproducibility module able to recognize the topology of the datasets over an RF algorithm and imputes missing values based on the inner data topology.

3   Notation, Related Work, and Methodology of Our Solution Approach

3.1   Notation

All notations and parameters used in the paper are summarized in Table 1.

Table 1. The notations and parameters in the paper are summarized below

Input notation		Framework parameters	
A	Input dataset in a matrix format	$nfld$	Number of folds cross validation
A'	Training set with completeness values	$nattr$	Number of attributes in the bagging
$A\emptyset$	Test set with missing values	$msiz$	Minimum decision tree size
a_m	Row vector of A'	tst	Type of dataset *(un/labelled)*
a_n	Attribute of A'	ndt	Number of trees
a_{mn}	Datum of a_m	$mdpt$	Maximum decision tree depth
$a_m\,\emptyset$	Row vector of $A\emptyset$		
$a_n\,\emptyset$	Attribute $A\emptyset$		
$a_{mn}\emptyset$	Datum of $a_m\emptyset$		

3.2   Related Work

Often, practitioners and researchers face the problem of missing values when treating either single or multiple datasets under ML, DM, or data fusion processes. Although many statistical methods can impute missing values, they can lead to undesirable outcomes, such as a) distortion of the original variance, b) distortion of covariance and correlation with other columns within the dataset, and c) over representation of the MFV when a large number of observations are missing [4]. Moreover, a simple deletion can considerably reduce the data for any multivariate analysis. It can be even worse if the loss affects a variable of interest in a small dataset. Aside from the statistics approach, researches on model-based analysis are proposing methods such as Principal Component Analysis (PCA) in the GLRM framework [10]. Their approach embeds the dataset with continuous, discrete, and binary values, into a low dimensional vector space. Due to its dimensionality reduction, it can be used to impute missing values. The limitation is that the data must corresponds to the ML model assumptions.

The GLRM uses quadratically regularized PCA, a model with normal distribution, varying between 0 and 1. With a different distribution, the GLRM model is no longer recommended. Petrozziello [17] presented a Distributed Neural Network (DNN), which deals with datasets composed of continuous values using minibatch stochastic gradient descent. The restriction is related to zero attribute values. Their model performs better with continuous values while struggling with the zeroes ones. The fuzzy similarity is used by [18] in a multidimensional time series context, in which the missing values are reconstructed with the average of the reference data. The limitations here are that the reconstruction of the missing values based on the weighted mean of the data in the training task leads to biases and the fact that the approach is restricted to time series data. Aggarwal [19] offered a conceptual reconstruction model based on PCA with continuous values. It depends on a dataset with a high correlation between attributes. In [11], researchers presented a method of using multiple datasets containing missing values. Their formulation is based on Bayes' rules. The proposed model limitations are (i) the model can only work with continuous values and, (ii) to impute missing values, the fusion process assumes i.i.d. of the data.

In a comparison among state-of-the-art methods (besides the statistical techniques), we found five model-based approaches: GLRM, PCA, DNN, Fuzzy Similarity, and Conceptual Reconstruction. In general, the significant limitations highlighted in the related work are: 1) data distribution assumptions, 2) deficiencies in dealing with heterogeneous data types, 3) the need for attributes correlation, 4) the assumption between the ML model and data type, and 5) i.i.d, which is rare in real-world datasets. The proposed framework overcomes the limitations since it presents a stacked ensemble framework based on the integration of an adapted RF, Jaccard index, and Bayes probability. In our proposal, we can tackle continuous, discrete, categorical, and binary data in the same dataset without an extra transformation. Our framework is not distribution dependent and can perform in linear and nonlinear datasets. Thus, our proposed ensemble ML framework can be employed in a broader scenario, configured in a single robust solution for missing values problems in a wide variety of datasets. The overall data type comparison is summarized in Table 2.

Table 2. Summary of model-based methods' coverage for different data types

–	Contin.	Discr.	Categ.	Bin.
GLRM	✓	✓	✗	✓
PCA	✓	✓	✗	✗
DNN	✓	✗	✗	✗
Fuzzy Similarity	✓	✗	✗	✗
Conceptual Reconstruction	✓	✗	✗	✗
Proposed Framework	✓	✓	✓	✓

3.3 Solution Approach

Our approach is to provide a broad and extensible framework that is capable of processing heterogeneous raw data. We also want to avoid making assumptions between data type versus ML model (e.g., the distribution of the data might be an assumption for some ML model), as previously discussed in Sect. 2. Indeed, we are considering the generalization of the of data representation. According to [14], the gain in this representation could be exponential according to its distribution or sparsity (e.g., a node in a decision tree, or one of the units in a restricted Boltzmann machine). Accordingly, a distributed representation of the data means that a large number of its possible subsets can be useful in representing the data in its totality. Then, we want to reuse different data samples without missing values to reconstruct the portion that suffers from those missing. Based on these objectives, we thus propose an innovative stacked ensemble framework to impute missing values based on data topology [15], comprised of three phases:

(1) adapted RF, (2) the Jaccard index, and (3) Bayes probability. The first phase is responsible for data ingestion and differentiating the rows with the completeness of information from those with missing values. In the given dataset A, in an m x n matrix format, m and n stand for the row vector and attribute, respectively. As depicted in this first phase in Fig. 2 (b), the entire dataset A consists of two portions of rows data: rows without missing values (complete rows) and rows that contain missing values. Then, the dataset A is subdivided into two parts: generating A' and $A\emptyset$. A' is a subset of the dataset A without missing values, and $A\emptyset$ is a subset of the dataset A that contains the rows with the missing values. After split the dataset A into A' and $A\emptyset$, these two subsets are inputs of the ensemble adapted RF method. We propose that, due to the possibility of combining multiple decision trees, the ensemble RF algorithm achieves better performance than an individual model. The RF produces better results because the trees are formed by different Bootstrap Aggregation (Bagging) (1) with the replacement of the row vectors a_m and sampling different dataset attributes a_n, as follows:

$$nattr = \log_2 a_n + 1 \qquad (1)$$

Bagging is the mechanism used to reduce the trees' variance as well as the correlation between the attributes, resulting in multiples trees with different variances. We made two important adaptations in our RF. The first was to carry on heterogeneous data types in the same dataset. The second was to store the data subset related to each node of the RF, according to the iterative top-down construction. Combining these two adaptations in the RF with the Jaccard index and Bayes probability allowed us to reach better results when compared with the benchmark (Statistical methods and GLRM framework) for heterogeneous datasets. We use the Gini Index (2) to weigh the proportion of the classes of each attribute and select the one with the smallest impurity.

$$\text{Gini}_i = 1 - \sum_{k=1}^{n} p_{i,k}^2 \qquad (2)$$

The value p refers to the proportionality of a_{mn} in the attribute a_n. It constitutes the learning phase. In the second phase, after the RF process, we inject the $A\emptyset$ portion of the data that contains missing values and greedily measures the similarity, using the Jaccard index (3), between each row with the missing values $a_m\emptyset$ and a_m rows of the subsets stored previously of each node for all trees in the RF process.

$$J(X,Y) = \frac{|a_m \cap a_m\emptyset|}{|a_m \cup a_m\emptyset|} \qquad (3)$$

The set of rows a_m with the maximum similarity to the row $a_m\emptyset$ is computed. This step will retain the set of rows a_m that have the maximum similarity to the row $a_m\emptyset$. The goal of this step is to filter the set of the rows in the dataset without missing values that are most similar to the row that has missing values. In the final step, the imputation in the missing values is applied based on Naive Bayes probability (4), and the value of the attribute a_{mn} is assigned to the attribute $a_{mn}\emptyset$. This step aims to define the value with the maximum probability to imputing the missing values. This framework with these three phases is different from other base-model approaches found in the literature. By applying the ensemble RF step, we have different data representations (trees with different variances) according to the bagging process in the learning phase. The Jaccard index works as a filter to select the rows without missing values with the maximum similarity to rows that contain missing values. By applying Bayes probability, we will have, among all the most similar rows, the attribute with the maximum probability value to impute the missing one. All these steps are illustrated in Fig. 2 (a).

$$P(a_{mn}\emptyset \mid a_{mn}) = \frac{P(a_{mn}\emptyset \mid a_{mn}) * P(a_{mn})}{P(a_{mn}\emptyset)} \tag{4}$$

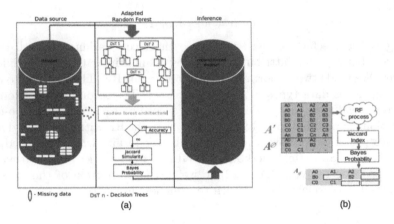

Fig. 2. (a) Conceptual framework of our model, (b) Synthetized algorithm mechanism.

4   Proposed Algorithm

In this section, we present in detail the proposed framework and discuss the typical complexity of working with heterogeneous data types in the same dataset.

Categorical Values: In real-world datasets, it is usual to find a large number of categorical columns. When these contain high cardinality (many different values), the traditional hot encoders found in the features engineering literature (e.g., one-hot encoding, count or frequency encoding) lead to a high number of new columns. That might affect the model due to the curse of dimensionality or a considerable loss of original information [4].

Algorithm 1: Ensemble learning for heterogeneous missing imputation

---

Input: A ', $A\emptyset$, $nfld$, $mdpt$, $msiz$, tst, ndt
1 **for** ndt_i in $[$ ndt $]$ **do**
2     $j \leftarrow 0$;
3   **while** $j <= nfld$ **do**
4        $Anfld \leftarrow$ **split** A 'in $nfld$ folds;
5   **compute** random forest $(Anfld, mdpt, msiz, tst, ndt\_i)$;
6   **update** $root$ $trees$ $\leftarrow$ random forest **results** ;
7   **if** tst $==$ $supervised$ **then**
8   **compute** $accuracy$ $\leftarrow$ metrics $(root$ $trees,$ $tst)$;
       end
    end
  end
9 **while** tst **do**
10     $v =$ **call fit** ($tst,$ $root$ $trees$) ;
11     $sim \leftarrow 1$;
12   **while** v **do**
13   **if** $Jaccard$ $Index(v,tst)$ $<sim$ **then**
14   **compute** $sim = v$;
       end
    end
  end
15 **if** $sim > 1$ **then**
16   **return** $argmax(sim)$
  else
17   **return** sim
  end

Regarding Numeric Values: It is often necessary in the presence of numeric values, to employ some feature transformation strategy, e.g., normalization, scaling, or binning, according to the numeric type (continuous or ordinal). This can affect the ML results. Thus, special attention is required in the preprocessing phase of the ML pipeline. Thereby, a summary of the frameworks reviewed in Sect. 3.2 and its inputs is presented in Table 2. The main steps of Algorithm 1 are as follows:

a) The subset A ', without missing values, and the subset $A\emptyset$ containing missing values are inputs of the framework.
b) Subsequently, the subset A ' is used to learn the RF algorithm. Besides the this subset, the algorithm receives the following parameters: $nfld$, $mdpt$, $msiz$, tst, and ndt (line 5).
c) We adapted the nodes and terminal leaves to them store the subset of the data (line 6).
d) In the inference phase of the framework, we evaluate the similarity between a_m and $a_m\emptyset$ applying the Jaccard index throughout the resulting trees (line 13).
e) In the next step, all a_m row vectors that satisfy the minimum Jaccard index is computed in line (14).

f) If we find a single row vector in the *root trees* with minimum Jaccard index, the value of this attribute will be the reference to impute the attribute $a_{mn}\emptyset$ (line 17).

g) Otherwise, if we found more than one row in the *root trees* with the minimum Jaccard index, the Bayes probability is used to define the most probable value to impute the missing line (16). In Fig. 2 (b), we synthesized the mechanism behind the algorithm.

5   Performance Evaluation

5.1   Principle of Comparison

We compare our framework's performance against statistical techniques and model-based methods of imputing missing values.

Comparison with Statistical Techniques: A set of techniques extensively used as a baseline in the statistical approach to continuous and discrete data types are mean and median imputation. MFV is a baseline to categorical and binary [24]. The mean, median, and MFV substitution replace the missing values with the local or global mean, median or MFV of the attribute of the set with missing values. These baselines are fast and easy to implement and can quickly scale with the dataset size. The drawback of these univariate imputation techniques is that it introduces bias and harms the statistical strength.

Comparison with Model-Based Approaches: The GLRM framework, which deals with heterogeneous data types, is implemented in relevant standard ML and database platforms, such as Spark, Python, Julia, and R. PCA is a well-known technique that minimizes the best rank-k by using the least-square. Due to the possibility of dimensionality reduction, it can be used to input missing values. Although PCA is a statistical technique, under the hood, its properties are widely used in many ML algorithms to deal with continuous values.

5.2   Datasets in the Case Study

Big data is defined regarding volume, variety, and velocity [9]. Currently, additional aspects like veracity, variability, complexity, and value have been proposed [5]. In this paper we are mainly exploring the variety and complexity of big data while experimenting with the IoT dataset, the images dataset, and datasets with heterogeneous data types. The volume is addressed in the experiment with a dataset that contains up to 310.000 rows; velocity is not the focus of the present paper. We chose five datasets to ensure the capability and extensibility of our framework when performing in datasets with different mixed data types, as follows:

1) The Air Quality (IoT) dataset of the city of Montreal available in the Smart City's open data platform. It contains the air quality of city, regarding five pollutants: Sulfur dioxide (SO2), Carbon monoxide (CO), Ozone (O3),

Nitrogen dioxide (NO2), and respirable fine particles (PM). It has 428 rows and five attributes (continuous, discrete, and categorical) in which, 128 rows 30% there are missing values in at least one attribute, it is a MCAR case.

2) Public Trees of the city is the dataset that catalog the trees in parks and streets of Montreal. From 316.070 rows, 220.027 69.61% represents the trees in the park and 96.043 30.39% represents the trees in the streets, a MNAR case. It has 22 categorical and discrete attributes.

3) The Synthetic dataset has 50 rows and 60 attributes (continuous, categorical, and binary) generated by the standard normal distribution.

4) The Fashion MNIST is a widespread image dataset with 60.000 rows in the training and 10.000 rows in the test set with 784 continuous attributes. We chose for evaluating the framework in the scenario with a unique data type.

5) The Titanic open dataset is a well-known in the ML community, with 891 rows and 12 attributes (discrete and categorical), it is a MAR case.

The specifications of these datasets are summarized in Table 3.

Table 3. Dataset details

Dataset	Data type	Dimensions	Type
Air quality	Contin./Categ.	428 × 5	Unlabelled
Public Trees	Contin./Categ.	316.070 × 22	Unlabelled
Synthetic Data	Contin./Categ./Bin.	50 × 50	Unlabelled
Fashion MNIST	Contin.	60.000 × 748	Labelled
Titanic	Contin./Categ.	780 × 22	Labelled

5.3   Evaluation Criteria

Criteria for Continuous and Discrete Values: We used the root mean square error (RMSE) (5), and coefficient of determination R2 (8) to evaluate the framework's performance compared to other techniques for the continuous and discrete values. The RMSE is a popular standard for measuring the error of the model for this type of attribute. There are continuous values with different scales in the datasets, due to it we use R2. It normalizes the scales usually, but not necessarily, from 0 to 1 and can assume negative values for example in non-linear functions. R2 is defined by the sum of the squared errors (SSE) divided by the total sum of the squares (SST).

$$\text{RMSE} = \sqrt{\frac{\sum_{i=1}^{n}(a_m - a_m\emptyset)^2}{n}} \tag{5}$$

$$\text{SSE} = \sum_{i=1}^{n}(a_m - a_m\emptyset)^2 \tag{6}$$

$$SST = \sum_{i=1}^{n} (a_m - mean(a_m\emptyset))^2 \qquad (7)$$

$$R2 = 1 - \frac{SSE}{SST} \qquad (8)$$

Criteria for Categorical, and Binary Values: For the categorical, and binary attributes we used the misclassification error (9) as the metric to evaluate the performance of our framework.

$$Miss.\ Error = \sum_{i=1}^{n} \frac{1, (a_m = a_m\emptyset); 0}{n} \qquad (9)$$

$$D_{KL}(P||Q) = \sum_{i} P(i) Log \frac{P(a_m)}{Q(a_m\emptyset)} \qquad (10)$$

$$D_{JS}(P||Q) = \frac{1}{2}D(P(a_m)||M) + \frac{1}{2}D(Q(a_m\emptyset)||M)\ where\ M = \frac{P(a_m)}{Q(a_m\emptyset)} \qquad (11)$$

6   Experiments and Results

In the learning phase, we performed the experiments before the imputation using the parameter ndt with 5, 10, and 15 trees for each dataset. In all of them, we ran five folds cross-validation partition, with 80% and 20% of subset A' as training and validation data, respectively. In the test phase, we ran the experiments 11 times, randomly removing the ground truth values of the attributes varying from 1% to 34%. Figure 3 shows the RMSE results for the 11 experiments in which attributes with missing values were increased from 1% to 34%. As expected,

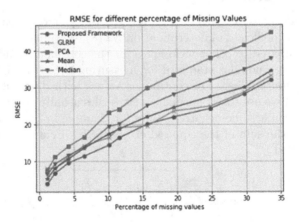

Fig. 3. RMSE for the experiments with continuous and discrete values in the datasets.

examining the RMSE growth for all techniques. However, the proposed framework is seen to outperform other players in all experiments. The RMSE of our framework varied from 3.81 to 32.01. The GLRM RMSE range varied from 8.13 to 33.17. It closely follows the proposed framework from the stage in which the percentage of attributes with missing values reaches 25%. The following are the mean, median, and PCA, respectively. Figure 4 shows the R2 metric for each technique. This second metric reinforces the advantage of our framework over the others, as it can be seen our framework reached a peak of 1.0 and a minimum of -1.04, with a mean of 0.47, the best imputation efficiency. Second is the GLRM, which ranges from 0.94 to -1.00 with a mean 0.36. The mean, median and PCA are 0.40, 0.23, and 0.08, respectively. This graph highligths the outliers present in all techniques. It is important to call attention to the high variance in the continuous values from different datasets. Considering all experiments, the average performance of the framework is 11.54% above the GLRM, 18.62% better than the mean, and 35.89% and 54.86% higher than median and PCA, respectively.

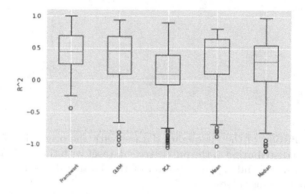

Fig. 4. R2 results for the experiments with continuous discrete values in the datasets.

Figure 5 offers the misclassification error for categorical and binary values. Clearly, the proposed framework achieves better results than the GLRM and MFV. The minimum result reached is 0.67 and the maximum is 0.85 for 2% and 17% of the missing values respectively. Due to the combination of our evaluation criteria (1-correct, 0-incorrect) and the low variance, all three methods struggled in the case of a small amount of missing values. However, even in this case our framework surpassed the other techniques. At the stage of 20% of missing values it is possible to see that our framework is fairly stable. Figure 6, shows the distributions of the categorical attributes in different datasets. As it can be seen, the attributes are non-normal distributed. In (a) the Kl (10) and Jensen-Shannon divergences (11) between the original distribution, our framework, the GLRM, and MFV are, respectively, 28.64, 37.85, and 69.27 for a categorical attribute in Air quality dataset. In (b), similarly, the Kl divergences for the three techniques

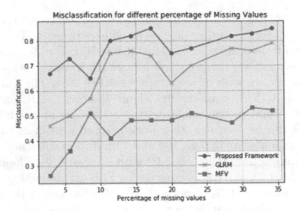

Fig. 5. Misclassification error for the experiments with categorical and binary values.

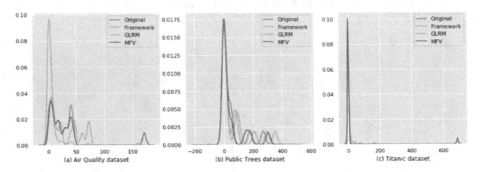

Fig. 6. (a) Distribution of the air quality dataset with the original 337 rows without missing values, (b) distribution of the public trees dataset with the original 220.027 rows without missing values, and (c) distribution of the Titanic dataset with the original 183 rows without missing values.

are 34.00, 47.16, and 56.49 for the Public trees. Finally, in (c), for a categorical attribute in the Titanic dataset, the Kl divergences are 480.45, 510.93, 686.78. The average performance of the framework in imputing categorical and binary type is 13.99% and 69.06% superior to the GLRM and MFV, respectively. In general, for the real-world dataset that suffers from the missing values problem, the goal is to retrieve the data in a manner that the retrieved data retain, as much as possible, the statistical power of the dataset. In Table 4, a summary of further statistical analyses, it is possible to analyse that the proposed framework provides the smallest divergence distribution, standard deviation, and mean, as compared to the other methods. Overall, the results demonstrate that our framework has obtained a small RMSE for continuous and discrete values, and it was able to preserve the similar distribution of imputing categorical and binary attributes compared to the benchmark. The interpretation of the result is due to the sequence of operation in each phase of the framework. In the initial phase, the

Table 4. Further statistical comparison for the distribution of categorical imputation data.

–	Dataset	KL Divergence	Jensen Shannon Divergence	Standard Deviation	
Framework	Air quality	28.64	0.093	24.88	30.5
	Public Trees	34.00	0.099	87.84	54.28
	Titanic	480.45	0.578	19.61	7.375
GLRM	Air quality	37.85	0.189	27.45	32.17
	Public Trees	47.16	0.105	83.12	52.12
	Titanic	510.93	0.773	35.04	13.45
MFV	Air quality	69.27	0.251	41.36	34.5
	Public Trees	56.49	0.162	84.2	47.5
	Titanic	686.78	0.902	146.08	33.71

RF produces a different subset of the data. In the second phase, by applying the Jaccard index, we select all subsets that match better with the subset where data is missing. Subsequently, we use Bayes probability to assign the most probable value to impute the missing data.

7 Conclusion

This work addresses the widespread missing values problem frequently found in real-world datasets. We proposed a comprehensive framework capable of addressing various data types, both in labelled and unlabelled datasets. The results obtained from the exhaustive experiments show that the proposed framework produces encouraging results and is competitive when compared to traditional statistical methods and other model-based frameworks. We experimented with framework using five datasets of different sizes and heterogeneous data types. The framework performed 11.54% superior to the GLRM, our benchmark in the model-based approach, and 18.62% better than the mean, the common statistical technique used. For categorical and binary types, our framework achieves performance that is 13.99% and 69.06% higher than the GLRM and MFV, respectively. Ultimately, it shows capacity to enhance the power of analysis and contributes to improving data quality by accurately reconstructing the dataset for the imputation of missing values appropriately. Hence, it can be employed in a wide range of datasets for supervised and unsupervised tasks. It is worthwhile to highlight that our approach addresses the categorical type as it is, without further transformation or encoding treatment. In the current and future scenarios in which the massive volume, variety, and velocity (big data pillars) of data are estimated, our framework is promising in addressing the missing values with respect to the volume and variety. Considering the future directions in which the present work can be extended, we can work on its implementation in the big data and ML platforms such as Spark and R.

Acknowledgments. The authors thank NSERC Strategic Program # 506319-17 for their financial support, and SARA at ÉTS for their writing support during this work, and the GLRM team for publishing their Synthetic dataset model, used as reference project at [16].

References

1. Mohan, K., Pearl, J.: Graphical models for processing missing data. arXiv:1801.03583, stat.ME (2018)
2. Azimi, I., Pahikkala, T., Rahmani, A.M., Niela-Vilén, H., Axelin, A., Liljeberg, P.: Missing data resilient decision-making for healthcare IoT through personalization: a case study on maternal health. Future Gener. Comput. Syst. **96**, 297–308 (2019). https://doi.org/10.1016/j.future.2019.02.015. ISSN 0167-739X
3. Hatem Ben Sta: Quality and the efficiency of data in "Smart-Cities". Future Gener. Comput. Syst. **74**, 409–416 (2017). https://doi.org/10.1016/j.future.2016.12.021
4. Schafer, L., Graham, J.W.: Missing data: our view of the state of the art. Psychol. Methods J. **7**, 147–177 (2002). https://doi.org/10.1037/1082-989X.7.2.147
5. Tan, Y., Zhang, C., Mao, Y., Qian, G.: Semantic presentation and fusion framework of unstructured data in smart cites. In: IEEE 10th Conference on Industrial Electronics and Applications (ICIEA), June 2015, pp. 897–901 (2015). https://doi.org/10.1109/ICIEA.2015.7334237
6. Cearly, D.W.: Top 10 strategic technology trends for 2019. Gartner Inc. and/or its affiliates. All rights reserved. PR575107 (2019)
7. Qin, X., Gu, Y.: Data fusion in the Internet of Things. Procedia Eng. **15**, 3023–3026 (2011). https://doi.org/10.1016/j.proeng.2011.08.567
8. Lau, B.P.L., et al.: A survey of data fusion in smart city applications. Inf. Fusion J. **52**, 357–374 (2019). https://doi.org/10.1016/j.inffus.2019.05.004. ISSN 1566-2535
9. Marjani, M., et al.: Big IoT data analytics: architecture, opportunities, and open research challenges. IEEE Access **5**, 5247–5261 (2017). https://doi.org/10.1109/ACCESS.2017.2689040. ISSN 2169-3536
10. Udell, M., Horn, C., Zadeh, R., Boyd, S.: Generalized low rank models. Found. Trends® Mach. Learn. **9**, 1–118 (2016). https://doi.org/10.1561/2200000055. ISSN 1935-8237
11. Housfater, A.S., Zhang, X.-P., Zhou, Y.: Nonlinear fusion of multiple sensors with missing data. In: IEEE International Conference on Acoustics Speech and Signal Processing Proceedings, vol. 4, p. IV, May 2006. https://doi.org/10.1109/ICASSP.2006.1661130
12. Sun, B., Saenko, K.: Correlation Alignment for Deep Domain Adaptation (2015)
13. Sun, B., Feng, J., Saenko, K.: Correlation alignment for unsupervised domain adaptation. arXiv:1612.01939, cs.CV (2016)
14. Bengio, Y., Courville, A., Vincent, P.: Representation learning: a review and new perspectives. arXiv:1206.5538, cs.LG (2012)
15. Bubenik, P.: Statistical topological data analysis using persistence landscapes. arXiv:1207.6437, math.AT (2012)
16. Udell, M., Horn, C., Zadeh, R., Boyd, S.: Generalized Low Rank Models (2016). https://github.com/powerscorinne/GLRM
17. Petrozziello, A., Jordanov, I., Sommeregger, C.: Distributed neural networks for missing big data imputation. In: 2018 International Joint Conference on Neural Networks (IJCNN), pp. 1–8, July 2018. https://doi.org/10.1109/IJCNN.2018.8489488

18. Baraldi, P., Di Maio, F., Genini, D., Zio, E.: Reconstruction of missing data in multidimensional time series by fuzzy similarity. Appl. Soft Comput. J. **26**, 1–9 (2015). https://doi.org/10.1016/j.asoc.2014.09.038. ISSN 1568-4946
19. Aggarwal, C.C., Parthasarathy, S.: Mining massively incomplete data sets by conceptual reconstruction. In: Proceedings of the Seventh ACM SIGKDD International Conference on Knowledge Discovery and Data Mining, KDD 2001, pp. 227–232. ACM, New York (2001). https://doi.org/10.1145/502512.502543. ISBN 1-58113-391-X
20. Albergante, L., et al.: Robust and scalable learning of data manifolds with complex topologies via ElPiGraph. CoRR Journal, vol. abs/1804.07580, August 2018. arxiv.org/abs/1804.07580
21. Bishop, C.M.: Model-based machine learning. Philos. Trans. R. Soc. Math. Phys. Eng. Sci. https://doi.org/10.1098/rsta.2012.0222
22. Zhou, Z.-H.: Ensemble Methods: Foundations and Algorithms. Chapman & Hall/CRC, Boca Raton (2012)
23. Geerts, F., Mecca, G., Papotti, P., Santoro, D.: Cleaning data with LLUNATIC. VLDB J. (2019). https://doi.org/10.1007/s00778-019-00586-5
24. Musil, C.M., Warner, C.B., Yobas, P.K., Jones, S.L.: A comparison of imputation techniques for handling missing data. West. J. Nurs. Res. **24**(7), 815–829 (2002)

Validating Goal-Oriented Hypotheses of Business Problems Using Machine Learning: An Exploratory Study of Customer Churn

Sam Supakkul[1(✉)], Robert Ahn[2], Ronaldo Gonçalves Junior[2],
Diana Villarreal[1], Liping Zhao[3], Tom Hill[2], and Lawrence Chung[2]

[1] NCR Corporation, Atlanta, GA, USA
{sam.supakkul,diana.villarreal}@ncr.com
[2] The University of Texas at Dallas, Richardson, TX, USA
{robert.sungsoo.ahn,ronaldo.goncalves,chung}@utdallas.edu,
tom.hill.fellow@gmail.com
[3] University of Manchester, Manchester, UK
liping.zhao@manchester.ac.uk

Abstract. Organizations are investing in Big Data and Machine Learning (ML) projects, but most of these projects are predicted to fail. A study shows that one of the biggest obstacles is the lack of understanding of how to use data analytics to improve business value. This paper presents *Metis*, a method for ensuring that business goals and the corresponding business problems are explicitly traceable to ML projects and where potential (i.e., hypothesized) complex problems can be properly validated before investing in costly solutions. Using this method, business goals are captured to provide context for hypothesizing business problems, which can be further refined into more detailed problems to identify features of data that are suitable for ML. A Supervised ML algorithm is then used to generate a prediction model that captures the underlying patterns and insights about the business problems in the data. An ML Explainability model is used to extract from the prediction model the individual features and their degree of contribution to each problem. The extracted weighted data feature are then fed back to the goal-oriented problem model to validate the most important business problems. Our experiment results show that *Metis* can detect the most influential problem when it was not apparent through data analysis. *Metis* is illustrated using a real-world customer churn (customer attrition) problem for a bank and a publicly available customer churn dataset.

1 Introduction

Data Analytics and Machine Learning (ML) technologies benefit from a continuous improvement cycle where large amounts of data are constantly being created. Organizations are investing in Big Data and ML projects, but most of these projects are predicted to fail [1]. A study may have suggested a possible

© Springer Nature Switzerland AG 2020
S. Nepal et al. (Eds.): BIGDATA 2020, LNCS 12402, pp. 144–158, 2020.
https://doi.org/10.1007/978-3-030-59612-5_11

reason which is the lack of understanding of how to use data analytics to improve business value [2]. This finding is a clear sign that stakeholders do not see the end-to-end relationship between important business goals and the emerging Big Data and ML technologies [3, 4].

Additionally, some business problems can only be hypothesized as they are difficult to validate using the traditional data analysis techniques. For example, applying data analysis on the customer churn dataset [5] during our experiment showed that there is no evidence to suggest that the customers who left the bank had a higher degree of dissatisfaction with many of the service qualities than those loyal customers.

Building on our previous approach, GOMA [6], this paper proposes *Metis*[1] to support goal-oriented hypothesis and validation of business problems. Three technical contributions are made in this paper, including 1) an ML-based approach to extracting an actual root cause hidden in the data to validate hypothesized business problems, 2) an ontology that more explicitly and formally describes the relevant modeling concepts related to business goals, problems and ML, and 3) a set of formalized validation rules for reasoning about problem hypothesis validation in a goal-oriented problem model.

The proposed approach is illustrated using a real-world banking customer churn problem, which was adapted from the example used in [6]. In the adopted case-study, a retail bank hired a company, specializing in data mining, to help address the churning problem by using insights from detailed transaction data in a newly installed powerful data warehouse [7]. The company hypothesized potential reasons to why the customers were canceling their accounts, and validated them with descriptive insights mined using a data classification technique. Since the actual dataset used by the consulting company was not available, we use a publicly available bank customer churn dataset [5] and reversed engineer to update the business problem hypotheses so that they are consistent with the dataset used. We use this example to demonstrate that *Metis* could be used to provide traceability between business problems and an ML solution, which can also reveal the insights about a root cause to a problem that may be difficult to discover using data analysis.

The rest of this paper is structured as follows. Section 2 presents the *Metis* method. Section 3 describes the experiment conducted and their results. Section 4 discusses related work along with our observed threats to validity and limitations. Finally, Sect. 5 summarizes the paper and future work.

2   Metis: A Goal-Oriented Problem Hypothesis Validation Method Using Machine Learning

To be able to hypothesize business problems and subsequently data features needed for developing an ML model, a good understanding of concepts in the domain in question is needed. In this section, we first present an example banking

---

[1] A Greek goddess that has been associated with prudence, wisdom, or wise counsel.

domain-specific ontology that underlies the customer churn problem that we use as a running example to illustrate Metis; we then describe the Metis domain-independent ontology used to support the modeling and validation of problem hypotheses in Metis; finally, we describe and illustrate the Metis process.

2.1 Domain-Specific Banking Ontology

This section describes an example of domain-specific ontology for the banking example, which can vary significantly depending on different organizations, processes, and so on. Figure 1 shows a typical set of banking concepts and their relationships. It is worth mentioning that this domain-specific ontology supports the understanding of the banking domain, and it does not represent a schema or model related to database design.

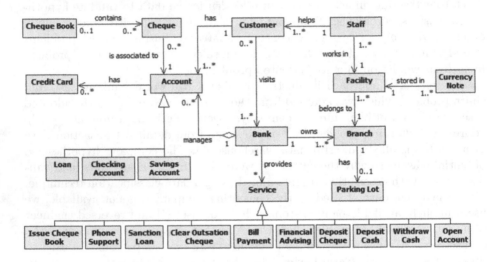

Fig. 1. Banking domain-specific ontology diagram

Some of the ontological concepts are briefly described here as examples. Banks provide numerous services, e.g., financial advising, cash withdrawal, and so on. It is important to study the qualitative aspect of these services in order to have a clear understanding of customer satisfaction. For example, customers may feel that there is not enough parking space, a lack of pleasant ambiance, no comfortable seating arrangements, lack of immediate attention, and so on. For the customer churn problem in the running example in this paper, the quality aspect of both facility and service related concepts are essential to generate hypotheses about problems.

2.2 The Metis Ontology

While modeling the mapping between a Goal-Oriented ontology and an ML-based ontology, completeness and soundness are two major concerns.

To completely and formally address these concerns for the *Metis* method, the following subsections describe the modeling concepts and the semantic reasoning formalization for the *Metis* ontology.

Modeling Concepts. A complete set of concepts and their relationships can be found in Fig. 2. To avoid omissions while mapping Goal-Orientation and ML, important concepts such as Problem, Hypothesis, and Machine Learning Model were explicitly represented. In addition, the ontology also comprehends concepts related to Big Data and Big Queries, and Features are derived from modeling concepts from a domain-dependant ontology. *Metis* is a domain-independent ontology that can be applied to a variety of domains. Section 2.1 describes a banking domain-specific ontology example.

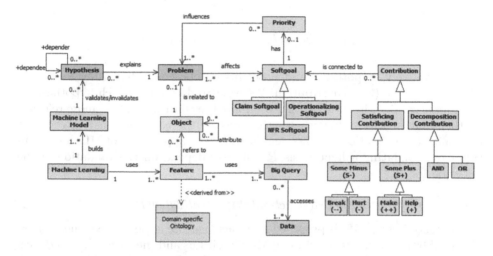

Fig. 2. Metis domain-independent ontology diagram

In a Goal-Oriented approach, problems represent undesirable situations, vulnerabilities, or threats to achieving stakeholders' goals. For the banking customer churn example in Fig. 4, *Customer Churn* is a problem that negatively impacts the goal *Retain existing customers*. Goals and problems can be further decomposed, and we can hypothesize what might contribute to their realization based on domain knowledge, statistics, and so on.

An acceptable representation of hypotheses and problems can be generated, but ultimately, we want to determine whether we can validate these hypotheses with respect to the problems in consideration. In this context, ML is used to build models to identify the importance of features such as *Immediate Attention*. By using the relevant features, it is possible to establish how to validate or invalidate hypotheses. For instance, in Fig. 4 we hypothesize that *Lack of immediate attention* has a S+/S− contribution to the problem *Poor Service*, which in turn contributes to *Customer Churn*.

Semantic Reasoning Formalization. An important aspect of contributions in the hypothesis validation process is the propagation of validations throughout the connected hypotheses, since a hypothesis might contribute to multiple other hypotheses. This validation process starts in the lowermost level of the hypotheses and propagates until we validate or invalidate problems. A formal definition may be described as follows:

Let $validated(P_n)$ be the proposition that the problem hypothesis P_n is validated, for $n \in \mathbb{Z}^+$. Let i be an arbitrary integer, where $i > 1$. For all $j \in \mathbb{Z}^+$, let $P_{i-1,j}$ be the jth problem hypothesis directly decomposed from P_i. Assuming this decomposition is of type OR/AND, the validation propagation can be represented by the following:

$$\left(\bigvee_j validated(P_{i-1,j}) \right) \rightarrow validated(P_i) \tag{1}$$

$$\left(\bigwedge_j validated(P_{i-1,j}) \right) \rightarrow validated(P_i) \tag{2}$$

Alternatively, hypotheses and problems can be connected using a positive (S+) or negative (S−) contribution. We want to determine which hypothesis in the source set (i.e., hypotheses that originate the contributions) is more relevant to the target set (i.e., hypotheses that receive the contributions) in order to maximize the validation insights generated by the application of ML models. In this case, validating a hypothesis P_i will now depend on the validation of the selection for P_i.

$$validated(selection(P_i)) \rightarrow validated(P_i) \tag{3}$$

By following the *Metis* proposed approach, feature importance values (I) are obtained from the results of running ML algorithms and they are associated with their respective contributions, i.e., a contribution from problem hypothesis P_s to P_t has an importance value $I_{s,t}$ and its weight and type may be determined by the following:

$$weight(P_s, P_t) = I_{s,t} \tag{4}$$

$$ctr\_type(I_{s,t}) = \begin{cases} S+ & \text{if } I_{s,t} \geq 0 \\ S- & \text{if } I_{s,t} < 0 \end{cases} \tag{5}$$

A source hypothesis has a score based on the weight of the targeted hypotheses and their respective contributions. The function $weight(P_t)$ describes the importance weight of a target hypothesis. Hence, the overall score for a source hypothesis P_s can be given by the utility function as follows:

$$score(P_s) = \left(\sum_{t=1}^{\#targets} weight(P_t) \times weight(P_s, P_t) \right) \tag{6}$$

After computing the scores for all source hypotheses using a utility function, the selection process may be carried out in a bottom-up approach [8]. We need to select the maximum value in the lowermost source hypothesis set to propagate that validation to the target hypothesis set. In other words, the selection process for a target is represented by choosing the source with highest score:

$$selection(P_t) = max\Big(score(P_s)\Big)_{s=1}^{\#sources} \qquad (7)$$

After the lowermost source hypothesis set is evaluated, we proceed to the next one until the selection process covers the entire set of hypotheses.

2.3   The Metis Process

The *Metis* process consists of four steps: 1) *Model business goals and problems*, 2) *Acquire data*, 3) *Detect feature importance*, and 4) *Validate hypotheses of business problems*. As shown in Fig. 3, *Step 1: Model business goals and problems* explicitly captures stakeholders' needs and obstacles as goals and problems using a goal-oriented modeling approach [9,10], where potential problems are posed as problem hypotheses to be validated. The outputs from this step are problem hypotheses in the context of business goals. *Step 2: Acquire data* derives data features from the business problems hypotheses to acquire the necessary data from external and/or internal sources, for instance using a customer survey or Big Data Spark SQL if the data are already available on-line. *Step 3: Detect feature importance* uses ML to learn patterns in the data to identify how problems are associated with the data features collectively. In addition, the output from this step includes Feature Importance that identifies the degree of each feature contributing to a problem. The final step, *Step 4: Validate hypotheses of business problems* uses the Feature Importance to validate the problem hypotheses modeled during Step 1.

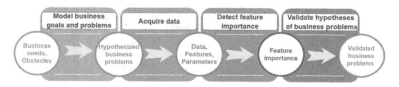

Fig. 3. The *Metis* process for validating goal-oriented hypotheses of business problems

Step 1: Model Business Goals and Problems. In this step, important business or stakeholders' needs are explicitly captured as Softgoals that can be further refined using AND or OR decomposition [11]. Using Fig. 4 as an example, at the highest organizational level, *Increased profitability* is a Softgoal to be achieved, which is refined using an AND decomposition to *Increased revenue* and *Increased profit margin* sub-goals, where the former is to be operationalized by *Increase customer base* strategic level goal. *Increase customer base* is then further AND-decomposed to more specific operationalizing goals of *Retain existing customers* and *Acquire new customers*.

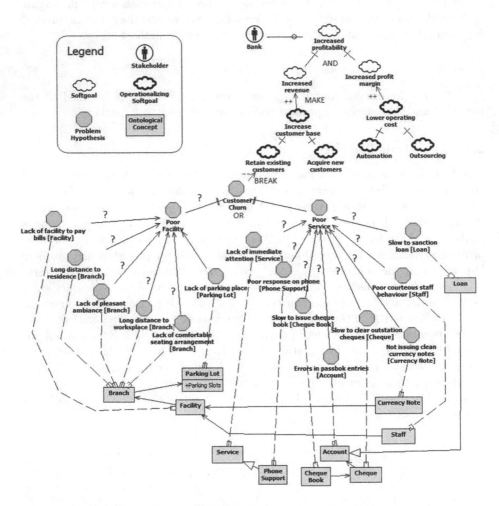

Fig. 4. Step 1: Model business goals and problems

Each lowest level goal is used as the context to identify potential problems that could hinder the goal achievement. The validity of each problem may be unknown at this point. Therefore, each problem is considered a target problem hypothesis to be validated by data. Similar to the goal refinement, each problem hypothesis may be further refined or realized by more specific problem hypotheses until they are low-level enough to identify the data features needed for data analysis or ML.

In this example, *Customer Churn* is a problem hypothesis that could *BREAK* (−) the *Retain existing customers* goal. *Customer Churn* is further refined using an OR-decomposition to *Poor Facility* or *Poor Service* sub-problem hypotheses, which are used to identify potential causing problem hypotheses. *Poor Facility* is hypothesized to be caused by *Long distance to residence, Lack of pleasant ambiance*, or other causes. Since each potential cause has not been validated whether it is indeed a contributing cause to the problem, the contribution link is labeled as unknown (depicted by a question mark).

Step 2: Acquire Data. This step examines the lowest level problem hypotheses to identify data features needed for data analysis or ML. Using Fig. 4 as an example, *Long distance to residence* and *Lack of pleasant ambience* may be used to identify *Distance to residence* and *Pleasant ambience* as the corresponding data features. The identified features are then used to build database queries or Big Data queries if the corresponding data are already available on-line. Otherwise, the required data features need to be acquired through other means, such as purchasing from a data provider, using a customer survey or generation from on-line sources [12].

An example of the acquired dataset is given in Fig. 5a, where F_1–F_5 represent all features and L corresponds to the *Churner* or *Non-churner* indicator associated with the satisfaction scores for F_1–F_5, provided by individual customers C_1–C_5. For example, customer C_1 expressed dissatisfaction with feature F_1 and F_3 with scores of 2 and 3 accordingly. On the other hand, he/she expressed satisfaction with F_2, F_4 and F_5 with scores of 8, 7, 8 accordingly. C_1 is noted by label 1 as a *Churner* customer in correlation with the given scores.

Step 3: Detect Feature Importance. The intuition for using ML is to encode the knowledge about the features hidden in the customer survey data, and then decode the knowledge representation to identify which feature is the true cause for the customer churn problem. To encode the feature knowledge, we use a Supervised ML algorithm with an assumption that an accurate prediction model represents the knowledge about features. To decode the influential features, we use a Model Explainability library [13] that was designed to explain how features contribute to the prediction outcomes.

Referring to Fig. 5b, this step splits the dataset into training and testing datasets. All features F_1–F_5 and label L are processed by one or more Supervised ML algorithm to obtain the most desirable prediction model M_p. To determine whether M_p has been sufficiently trained to recognize the general patterns in the training dataset, it is measured on how accurate it can predict label L in the testing dataset. The accuracy is represented by an accuracy metric A_1, which is based on the differences between predicted label L' and actual label L, where L' is generated from F_1–F_5 in the testing dataset.

Once an accurate model M_p is obtained, it is processed by an ML Explainability algorithm to produce an Explainability model M_e, which is in turn used to detect feature importance I_1–I_5, where I_1 contains two pieces of information: sign and weight of the contribution F_1 makes towards the label L' as predicted M_p. The sign of the value indicates whether the corresponding feature helps or hurts towards the predicted label, while the weight represents the amount of influence the feature has. Similarly, I_2 and I_3 represent the feature importance of F_2 and F_3 respectively. By having the highest value among all feature importance values, F_1 is considered the most influential feature, followed by F_2 and F_3 in the context of testing dataset.

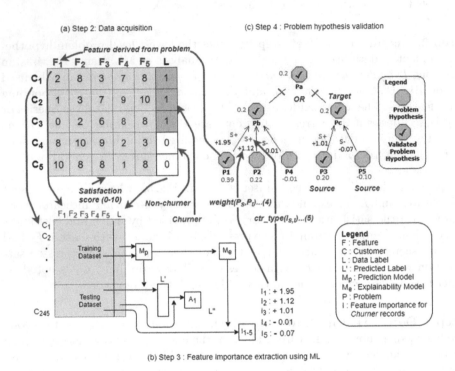

Fig. 5. Step 2, 3, 4 of the *Metis* process

Step 4: Validate Hypotheses of Business Problems. Referring to Fig. 5c, this step uses the feature importance values produced by the Explainability model M_e to validate problem hypotheses in the goal-problem model created in step 1, one parent-child problem set at a time in a bottom-up approach, using the quantitative and qualitative semantic reasoning formalization, as described in Sect. 2.2.

Using P_b - (P_1, P_2, P_3) parent-child set as an example, the contribution link between each parent-child pair is updated by applying Formula 4 and 5 against the corresponding feature importance value where (P_1, P_2, P_3) and Pb are considered *sources* and *target* in the formulas respectively. In this example, the contribution type $ctr\_type(P_1, P_b)$ is assigned with $S+$ by Formula 5 with feature importance I_1 with value $+1.95$ as a function parameter. I_1 is used since the corresponding F_1 was defined based on problem P_1 in step 2. To complete the contribution update, the weight of contribution is assigned with 1.95 by

Formula 4. Other contribution links with the same parent are updated in a similar fashion. Then, P_1 is selected among P_1, P_2 and P_3 by Formula 7 to be a validated problem hypothesis since it is the most influential cause for problem P_b. After P_1 is quantitatively selected based on scores, P_b is qualitatively validated by Formula 3. Then, P_a can be qualitatively validated by Formula 1.

3   Experiment and Results

The analysis of customer feedback information may be beneficial to discerning customer satisfaction for the quality of important services. To this end, a publicly available dataset [5] acquired by Step 2 in the *Metis* process is analyzed in this section. This dataset contains typical customer information such as age and occupation. In addition, the dataset contains feedback information regarding certain banking service-related features (e.g., Immediate Attention) and facility-related features (e.g., Pleasant Ambiance). A customer can score each of these features from 0 to 10 (least to most satisfied). In this context, scores of 4 or less are used to describe some degree of dissatisfaction, an assumption that something might go wrong in a business operation, i.e, problem hypotheses. The next section describes an analysis of these problem hypotheses.

3.1   Dataset Analysis

Some examples of problem hypotheses are shown in Fig. 4, which includes *Long distance to residence, Lack of immediate attention,* and *Lack of pleasant ambiance.* Figure 6 shows the customer dissatisfaction with respect to these 3 features out of the 20 available features. From the total of customers that believe there is a long distance to their residence, 35% deserted the bank (churner) and 65% remained loyal (non-churner). Assessing this feature by occupation, notice that most unsatisfied customers are from the professional occupation, followed by private service and government service. Together, these three occupations represent 75% of customers unsatisfied with distance from residence.

From the customers that identified a lack of immediate attention, more than half are young customers (40 years old or younger). Analyzing pleasant ambiance by occupation, we can see that customers from the business occupation and the private service complained the most. In an overall assessment for customer dissatisfaction by loyalty, it is possible to notice that most customers remained loyal regardless of the problem hypotheses under consideration. Even though we are able to extract insights from the dataset, ultimately, there is no evidence of why customer deserted. For this purpose, Sect. 3.2 demonstrates results of using ML that can potential provide some evidence.

3.2   Prediction Models

To encode and represent knowledge about feature contributions using Supervised ML, we experimented with several ML algorithms, including Linear Regression,

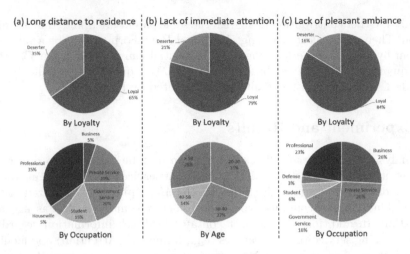

Fig. 6. Analysis of customer dissatisfaction (score less than 5 in a 0 to 10 scale) for the features distance from residence, immediate attention, and pleasant ambiance

Support Vector Machine, Decision Tree, Random Forest, and XGBoost Classifier, among which XGBoost showed the highest accuracy rate in our experiment. Due to space limitation, only the results from XGBoost are discussed in this section.

The ML segments of the experiment were conducted using Python language and scikit-learn open-source ML libraries [14]. The dataset used for the experiment was a public banking customer churn dataset [5]. After data cleansing, 67% of the data (164 records) were used for model training and the 33% (81 records) for testing. Data features used included the customer responses to the survey questions, such as *Pleasant Ambiance, Comfortable Seating, Immediate Attention, Good Response On Phone* and others, in the scale of 0–10, excluding customer information, such as age and occupation, that were used separately for data analysis as reported in Sect. 3.1. The resulting prediction model showed an accuracy of 84% (F1 score) on the test dataset, which was better than other ML algorithms in our experiments. The modest accuracy rate was probably due to the small and highly unbalanced dataset that required a data pre-processing step that further reduced the dataset size.

3.3 Explainability Model

To extract feature contribution information from the resulting prediction model, we used SHAP (SHapley Additive exPlanations) [13], an Explainability library that uses a game theoretic approach to explain the output of many ML models. It connects optimal credit allocation with local explanations using the classic Shapley values from game theory and their related extensions.

Figure 7 is a Force Plot produced from a SHAP model (M_p in Fig. 5b) created from the most accurate XGBoost prediction model (M_e in Fig. 5b). It gives a visual representation of the influence each feature has towards the final output

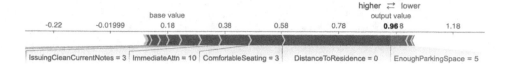

Fig. 7. Features importance for one churner's responses

value of 0.96. In this plot, the base value 0.18 is the average prediction value without any influences from the features, while output value of 0.96 is the output from the prediction model where 1 represents a churner customer. The influences of features are represented by the direction towards the output value and width of the corresponding arrow blocks. Here, *DistanceToResidence* feature has the most influence in increasing the output value away from the base value towards the final output value, which is consistent with the score of 0 (least satisfaction) given by the customer. On the other hand, *EnoughParkingSpace* has the most influence in the opposite direction, decreasing the value away from the final output value, which seems consistent with the satisfaction score of 5 (neutral satisfaction) given by the customer. It is interesting to note that *ImmediateAttn* with the value of 10 (most satisfaction) was seen as an influence towards the customer's churner decision. SHAP does explain this counter intuitive result.

Figure 8a plots individual SHAP values for all features and all churner customers. Each dot represents a SHAP value that a feature has in support of increasing the output value towards 1 (Churner label in Fig. 5a). Visually, it is clear that *DistanceToResidence* has higher positive SHAP values than other features. For this experiment, the more positive SHAP values a feature has, the more influence it has towards the prediction outcome. This is supported by Fig. 8b where *DistanceToResidence* has the highest total(sum) SHAP value.

3.4   Validating Problem Hypotheses

By following Step 4 of the *Metis* process (Sect. 2.3), we applied Formula 4 and 5 against the sum SHAP value for the respective feature (see Fig. 8b), which led to the validation of *Long distance to residence* problem hypothesis against other features having Poor Facility as the common parent problem hypothesis. Then, Formula 3 was applied to validate Poor Service problem hypothesis. Subsequently, Formula 1 was applied to validate *Customer Churn* problem hypothesis. The resulting goal-problem model is shown in Fig. 9, with check marks to reflect the validation status.

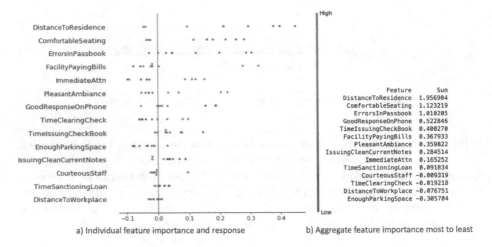

a) Individual feature importance and response

b) Aggregate feature importance most to least

Fig. 8. Features importance for all churners' responses

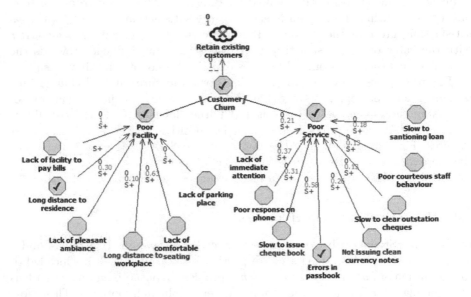

Fig. 9. Validated customer churn problems

4   Related Work and Discussion

We believe this paper is one of the first to propose an end-to-end, explicit and formal approach that provides traceability between business goals and ML. Most data mining and ML projects in practice are often based on informal identification of low-level problems [7] that may not have clear relationships with higher level goals. *Metis* allows ML solutions to be traceable to business at the highest level of business goals and corresponding problems. Using data to validate goal-oriented models have been proposed in [15] using questionnaire and the statistical hypothesis testing to validate different model elements (e.g, actors, goals, resources) and their relationships (e.g, depends, make, hurt). The statistical method is widely accepted but has been criticized to be difficult to understand [16] and impractical to find evidence in the real-world for some hypothesis to test the null hypothesis [17]. This is especially true in the data-rich Big Data environment, where it is difficult to find evidence for both hypothesis and null hypothesis in the available business data. ML allows organizations to utilize the existing data for hypothesis validation that is grounded by the model accuracy.

Threats to Validity and Limitations. Regarding threats to internal validity, the dataset used in the experiment was highly relevant to the customer churn problem, but it was a small dataset (i.e., 245 records), which could lead to biased results. To reduce this bias, training and testing data were randomly selected and tested with stratification. We also ran several ML algorithms but got similar results. For threats to external validity, as we only applied our approach to a customer churn case, the approach may be too early to be generalized. More experimentation for different domains and datasets is needed.

This paper has presented a promising initial result with some limitations, including 1) inter-feature AND and OR relationships are not currently supported, 2) it is currently unclear whether the result would be consistent across other ML algorithms and model explainability libraries.

5   Conclusions

This paper presents, *Metis*, a novel approach that uses ML to validating hypotheses of business problems that are captured in the context of business goals. *Metis* uses Supervised ML and Model Explainability algorithms to detect feature importance information from the data. Our initial experiment results showed that *Metis* was able to detect the most influential problem root cause when it was not apparent through data analysis. The most influential root cause was then used to validate higher level problem hypotheses using the provided formalization.

Future work to address the identified threats to validity and limitations include 1) conducting additional experiments with larger datasets, 2) testing with additional ML algorithms and explainability libraries, 3) investigating solutions for encoding AND/OR relationships in the datasets for model training or exploring ML algorithms internally for an ability to extract the relationships if captured by the algorithms.

References

1. Asay, M.: 85% of big data projects fail, but your developers can help yours succeed. Techrepublic (2017)

2. LaValle, S., Lesser, E., Shockley, R., Hopkins, M.S., Kruschwitz, N.: Big data, analytics and the path from insights to value. MIT Sloan Manag. Rev. **52**(2), 21–32 (2011)
3. NewVantage Partners LLC: Big data and AI executive survey 2020: data-driven business transformation connecting data/AI investment to business outcomes (2020)
4. Zhou, L., Pan, S., Wang, J., Vasilakos, A.V.: Machine learning on big data: opportunities and challenges. Neurocomputing **237**, 350–361 (2017)
5. Tatter, P.: gpk: 100 data sets for statistics education. R package v. 1.0. (2013)
6. Supakkul, S., Zhao, L., Chung, L.: GOMA: supporting big data analytics with a goal-oriented approach. In: 2016 IEEE International Congress on Big Data (Big-Data Congress), pp. 149–156, June 2016
7. Berry, M.J., Linoff, G.S.: Data Mining Techniques: for Marketing, Sales, and Customer Relationship Management. Wiley, Hoboken (2004)
8. Supakkul, S.: Capturing, organizing, and reusing knowledge of non-functional requirements: an NFR pattern approach. Ph.D. thesis, University of Texas at Dallas (2010)
9. Chung, L., Nixon, B.A., Yu, E., Mylopoulos, J.: Non-Functional Requirements in Software Engineering, vol. 5. Springer, Cham (2012)
10. Supakkul, S., Chung, L.: Extending problem frames to deal with stakeholder problems: an agent-and goal-oriented approach. In: Proceedings of the 2009 ACM Symposium on Applied Computing, pp. 389–394 (2009)
11. Mylopoulos, J., Chung, L., Nixon, B.: Representing and using nonfunctional requirements: a process-oriented approach. IEEE Trans. Softw. Eng. **18**(6), 483–497 (1992)
12. Wang, W., Chen, L., Thirunarayan, K., Sheth, A.P.: Harnessing Twitter 'big data' for automatic emotion identification. In: 2012 International Conference on Privacy, Security, Risk and Trust and 2012 International Conference on Social Computing, pp. 587–592. IEEE (2012)
13. Lundberg, S.M., et al.: From local explanations to global understanding with explainable AI for trees. Nat. Mach. Intell. **2**(1), 56–67 (2020)
14. Pedregosa, F., et al.: Scikit-learn: machine learning in python. J. Mach. Learn. Res. **12**, 2825–2830 (2011)
15. Hassine, J., Amyot, D.: A questionnaire-based survey methodology for systematically validating goal-oriented models. Requirements Eng. **21**(2), 285–308 (2015). https://doi.org/10.1007/s00766-015-0221-7
16. Cohen, J.: The earth is round (p<.05). Am. Psychol. **49**, 997–1003 (1994)
17. Berkson, J.: Tests of significance considered as evidence. J. Am. Stat. Assoc. **37**(219), 325–335 (1942)

The Collaborative Influence of Multiple Interactions on Successive POI Recommendation

Nan Wang[1,2], Yong Liu[1], Peiyao Han[1], Xiaokun Li[2(✉)],
and Jinbao Li[1(✉)]

[1] School of Computer Science and Technology, Heilongjiang University,
Harbin 150080, China
lijbsir@126.com
[2] Postdoctoral Program of Heilongjiang Hengxun Technology Co., Ltd.,
Harbin 150090, China
xiaokun_li_hx_hlju@163.com

Abstract. With the rapid development of social networks, users hope to obtain more accurate and personalized services. In general, POI recommendation often uses the historical behaviors of a user to recommend the top N POIs and rarely consider the current state of the user. Unlike POI recommendation, successive POI recommendation is more sensitive to user preferences and changes in time and space. In order to alleviate the data sparsity, we make full use of the interaction of time, space and user interest preferences, and propose a successive POI recommendation model called UTeSp. The UTeSp model uses the collaborative influence of multiple interactions to build a model, which can well adapt to the needs of users at different times and different locations. And it can change dynamically. Furthermore, we associate the user's inherent interest preference with the user's friend's influence on the target user, and propose a user-level interest preference based on attention mechanism, which can obtain more accurate user preference results. In addition, a novel TDP_HC algorithm is designed to segment time dynamically. Based on the partial order relationship, we propose two interpretable methods to enhance the learning ability of the model. The two methods can be used in other similar successive POI recommendation models. Experimental results show that the F1-score of UTeSp model on the two real datasets is better than that of several mainstream successive POI recommendation models, and the two partial order methods also show the effectiveness of our model.

Keywords: POI · Attention mechanism · Successive POI recommendation · Time dynamic partition

1 Introduction

In Location-Based Social Networks (LBSNs), users can enjoy numerous social network services. POI recommendation utilizes users' historical behaviors to recommend the right POIs [1, 2]. With the maturity of social networks, people pay more attention to

© Springer Nature Switzerland AG 2020
S. Nepal et al. (Eds.): BIGDATA 2020, LNCS 12402, pp. 159–174, 2020.
https://doi.org/10.1007/978-3-030-59612-5_12

personalized recommendation results, such as the next recommendation based on the current user's location or different time and so on. It is a kind of successive POI recommendation.

Different from the traditional recommendation, successive POI recommendation is based on the continuous historical behaviors of users [3]. It enables users to obtain the next POI location satisfying users when them are on the current POI location [4]. For example, when a user is shopping in a mall, he may be more interested in recommending the discount information of the nearby mall. However, it's also full of more challenges. Firstly, a user's continuous check-in behavior is more sparse than discontinuous behavior. How to get effective information from sparse data is a key problem. Secondly, the user's interests will vary in different scenarios, such as being influenced by friends, choosing different POI on weekdays and non-weekdays, and the distance between POIs, which will affect the recommendation results. How to meet different users' individual needs is also an important issue. Thirdly, the successive POI recommendation is based on the current location, its modeling process is more complex, so the model learning is relatively difficult. The existing successive POI recommendation did not make full use of the auxiliary information of context in social network, and did not get more abundant user's interest preference information, which makes it difficult to meet the user's personalized needs.

In this paper, we try to use a lot of information to build a complex interaction process, and propose a model based on user-level attention to effectively solve successive POI recommendation problem. This model has the ability to obtain the user's personalized interest by learning various interaction relationships. In addition, we propose two interpretable and personalized learning methods to learn the model. The main contributions of this paper are as follows:

(1) Based on multiple information in social network, combined with the multiple interaction among user's interest preference, dynamic time partition and variable spatial locations, a successive POI recommendation model named UTeSp model is proposed. (2) We use the attention mechanism to combine the user's internal interest preference and the influence of the user's friends. And we also design a time dynamic partition algorithm based on hierarchical clustering to make the temporal segmentation more reasonable. All of these further ensure that the users' preferences can change dynamically with different needs of users. (3) We propose two interpretable learning methods based on locational partial order and temporal partial order to establish the optimization goal, which can effectively learn the model. And the two methods can also be applied to other successive POI recommendation models. (4) The experiment results show that the performance of the UTeSp model is better than many mainstream models.

2 Related Work

Rendle et al. [5] proposed the earliest successive recommendation model, which called FPMC. They used Markov chain and matrix factorization methods to predict the next POI's location. Cheng et al. [6] extended their work, they used a combination of region partitioning and Markov chain for successive POI recommendation. They took the user's current location as the query POI, and the next location as the target POI, and propose a new FPMC-LR model based on matrix decomposition method. Liu et al. [7] used the POI category attribute to establish a transfer model to predict the next POI's location. Because the accuracy of the model depended heavily on the results of POI classification, the performance of the model was general. Feng et al. [8] first defined a successive new POI recommendation problem. They not only proposed a method of personalized metric embedding to model the user's check-in sequence, but also considered the impact of geographic factors.

From a context perspective, Zhao et al. [9] began to study the influence of time-space factors on successive POI recommendation, and proposed a time-sensitive model. The model skill-fully designed a time indexing mechanism to represent the timestamp as an integer value, which got a meaningful result. He et al. [10] proposed a successive POI recommendation model based on Bayesian sorting method of category preference, however the influence of other factors was ignored. Wu et al. [11] used graph convolutional network to build a social influence model for social recommendation. Wang et al. [12] utilize social information to capture user social exposures rather than user preferences for recommendation. Rahmani et al. [13] and Zhang et al. [14] use geographical influences to fuse the technology of matrix factorization for POI recommendation. Chen et al. [15] also used attention network to combine with collaborative filtering technology to recommend multimedia information. Although there have been many such works in the past, most of them were limited in traditional POI recommendation, and little attention was paid to the interaction modeling of multiple information based on attention on successive POI recommendation. In this paper, we propose a personalized recommendation based on user level attention.

3 UTeSp Model

Next, we define the successive POI recommendation model and describe the whole UTeSp model's modeling process in detail.

Definition 1 (Successive POI Recommendation)
For a given user u ($u \in U$), according to his historical check-in behaviors, he is at POI location l_{cur} at time t, and can be at POI location l_{next} at time $t + 1$. The successive POI recommendation model is to find the top N unvisited POI locations that meet the above conditions. We also define some important notations in the paper, as shown in Table 1.

3.1 User-Level Attention Modeling

In real social networks, users are more likely to accept suggestions from friends or close people [16]. In other words, a user's interest preferences may be influenced by his friends. However, for different friends of a user, each friend has different influence on the user. Based on this, we propose a new user interest preference that utilizes a user-level attention mechanism. First of all, we believe that users' interest preferences can be divided into internal interest preferences and external interest preferences. Secondly, we regard the user's inherent interest preference as the user's internal interest preference, and the user's friend's influence on the user as the user's external interest preference. Thirdly, we use attention mechanism to gain the influence of different users' friends. Finally, we combine the user's inherent interest preferences with the user's friend's influence results to form a new user's interest preferences. This result makes the latent vector information of the user more abundant and can get better user preference information. See formula (1) for details.

$$U_i = U_i' + \sum_{j \in F(i)} \frac{a_{ij} U_j}{|F(i)|} \tag{1}$$

In formula (2), the user preference latent vector U_i of user i is expressed in two parts, U_i' represents the internal latent vector of user i, and $a_{ij} U_j$ represents the result of the influence of friend j on user i. Here, U_j is the latent vector of interest preference of friend j, and a_{ij} is the attention weight. $F(i) = \{j| j: s_{ij} = 1\}$ is the friends set of user i, and s_{ij} is the social relationship matrix of user. When $s_{ij} = 1$, it means that user i and j are friends; otherwise, $s_{ij} = 0$.

Table 1. Notations

Notations	Descriptions
a_{ij}	The attention weight of user i to his friend j
w_{ij}	The similarity measure between user i and user j
d_i	The i-th check-in record
U, T, L	The sets of users, timestamps and POI's locations
l_{cur}, l_{next}	The current POI's location and the next POI's location
$\zeta_{lcur}(l)$	The possibility of the user from the current POI's location to the next POI's location to be visited
$(u, t, l_{cur}, l_{next})$	A four tuple of the check-in record that user u is at l_{cur} at time t, and l_{next} at time $t + 1$
$h(u, t, l_{cur}, l_{next})$	The evaluation function of the model, indicating the evaluation result of possibility that user u is at l_{cur} at time t and l_{next} at the next time
$U_u, L_{cur}, L_{next}, T_t$	The feature vectors of the user u, the current POI's location l_{cur}, the next POI's location l_{next}, the time segmentation t
O^L_{UTeSp}, O^T_{UTeSp}	The successive recommendation optimization targets based on locational partial order relationship and time partial order relationship

We use a nonlinear function *ReLU* with transform matrix w_0 to get a new user's friend vector U_j. Where, b_0 is a bias vector, U_j' represents the internal latent vector of user j. See formula (2) for details.

$$U_j = ReLU(w_0 * U_j' + b_0) \tag{2}$$

And the attention weight a_{ij} is obtained by normalizing user friend vector with user similarity using the softmax function. We use the friend vector of user i with user similarity to obtain the attention weight a_{ij} of friend j which has different influence on user i. As shown in Formula (3). Where, w_{ij} is the result of the similarity measure between i and j. Next, we discuss two different cases of w_{ij}.

$$a_{ij} = \frac{\exp(w_{ij}U_j)}{\sum_{j \in F(i)} (w_{ij}U_j)} \tag{3}$$

(1) Based on the similarity of common friends

When q is a common friend of user i and j, the common friend similarity measure w_{ij}^a can be defined using cosine similarity. Where s_{iq} (or s_{jq}) is an element in the social relationship matrix. User i and user q have an edge, $s_{iq} = 1$ (or $s_{jq} = 1$), otherwise $s_{iq} = 0$ (or $s_{jq} = 0$).

$$w_{ij}^a = \frac{\sum_{q=1}^{|U|} s_{iq}s_{jq}}{\sqrt{\sum_{q=1}^{|U|} (s_{iq})^2}\sqrt{\sum_{q=1}^{|U|} (s_{jq})^2}} \tag{4}$$

(2) Based on the similarity of common check-in behaviors

In the above similarity measure, we ignore a problem: when user i and user j are friends, but there is no common friend, the value of similarity is 0, which means that the influence between friends does not exist at all. In fact, they not only affect each other, but also play a very important role. Therefore, we adopt the idea of the common check-ins to solve this problem. We use the ratio of the locations that two users have visited together to all the locations they have visited to express the similarity w_{ij}^b based on the common check-in behaviors.

$$w_{ij}^b = \frac{K(i) \cap K(j)}{K(i) \cup K(j)} \tag{5}$$

$K(i)$, $K(j)$ represent the set of all POI locations visited by user i and user j, respectively. Finally, we define the similarity between user i and his friend j using the idea of weighted average. It considers both common friends and common check-ins.

$$w_{ij} = \beta * w_{ij}^a + (1 - \beta) * w_{ij}^b \tag{6}$$

Where β is the weight value of users i and j based on the similarity of common friends.

3.2 Dynamic Temporal Modeling

In the past, time sensitive recommendations always divide a day or a week manually, and all user follow the same distribution. In fact, due to the different living habits and hobbies of each person, the law of human activities also has a strong personality. Therefore, for different users, the historical behavior information of users in different time periods is not same. Accordingly, we conclude that every user should have different time partition criteria to determine his behavior rule. We propose a Time Dynamic Partition algorithm based on Hierarchical Clustering (TDP_HC algorithm). The method of hierarchical clustering is used to select two adjacent time segments with the smallest time interval, merge into a new time segment, and iteratively update until the entire check-in record is divided into the required number of segments s by time (In this paper, $ns = 8$). The TDP_HC algorithm as follows:

TDP_HC Algorithm

Input: $D=\{d_i=(u,t,l)|(u,t,l) \in U \times T \times L\}$, ns //ns:the number of segmentations, d_i represents a check-in record that user u visited POI l at time t; U, T, L represent different sets of users, time periods, POI locations

Output: TS

   1: $TS=\phi$;   //TS is the time segmentations' set of time dynamic partition

   2: For each $d_i \in D$ do

   3: $s_i=(d_i.start, d_i.end, d_i.avg)$;$TS=TS \cup \{s_i\}$;   // each record d_i is placed in the set TS in the form of (the start time, the end time, the average time) as the elements of the TS; .avg is the average time of the segment as the current segment time

   4: While $|TS|>ns$ do

   5:     $min_{dist}=+\infty$;   // the minimum time interval

   6:     For each $s_i \in TS$ do

   7:         If $| s_i.avg - s_{i+1}.avg |< min_{dist}$ then

   8:             $min_{dist} = | s_i.avg - s_{i+1}.avg |$; $s_a=s_i,s_b=s_{i+1}$;   // Find the minimum time interval, and record the location of subsegments

   9:         $TS=TS-\{s_a, s_b\}$;

   10:         $s_c=(min(s_a.start,s_b.start),max(s_a.end,s_b.end),(s_a.avg+s_b.avg)/2)$;   //$s_c$: A new subsegment

   12:     $TS=TS \cup \{s_c\}$;   //merge s_c into TS

   13: Return TS

Since the initial size of the TS is $|D|$, the time complexity of the while loop in line 4 is $O(|D|)$, the time complexity of the for loop in line 6 is also $O(|D|)$, and the time complexity of the entire algorithm is $O(|D|^2)$.

3.3 Variable Spatial Modeling

In fact, most users in the real world will choose a closer location to the current location for the next visit and only a few people will choose to go to the place far away from them, so human check-in behavior will generally follow the power law distribution

[17]. Inspired by Yuan et al. [18], we add the influence of geographical information in the successive POI recommendation model.

The power law distribution $p_{ro} = c|k|^{-\rho}$ is used to model the user from one location to another. Where $|k|$ is the distance between the user's current location and the next to be visited, p_{ro} indicates the willingness of the user to go to the POI that $|k|$ km away from themselves, c, ρ are parameters of the power law distribution function and can be obtained by maximum likelihood estimation. Because the linear function above can be learned by the least-squares regression method, for the current location l_{cur} and the next location l_{next} to be visited, we can use the power law distribution of the user's check-in behavior model the possibility of the user from the current location to the next to be visited. The possibility is expressed in $\zeta_{lcur}(l)$.

$$\zeta_{lcur}(l) = \frac{c|l_{next} - l_{cur}|^{-\rho}}{\sum_{l_{next} \in L, l_{next} \neq l_{cur}} c|l_{next} - l_{cur}|^{-\rho}} \tag{7}$$

Where l_{cur} and l_{next} respectively represent the current POI's location and the next POI to be selected. All POIs' locations are represented by geographical coordinates (including longitude and latitude). When the distance $|l_{next} - l_{cur}|$ is small, $\zeta_{lcur}(l)$ is relatively large, indicating that users are more willing to choose a POI location close to themselves to visit. On the contrary, $\zeta_{lcur}(l)$ is relatively small. Therefore, we set $\zeta_{lcur}(l)$ as the weight value to describe the relationship between the current location and the next location. It can well express the influence of spatial information between two different locations. It may be a good choice.

3.4 Model Definition

We use the comprehensive influence of user interest perference, time segmentation, and spatial information to build UTeSp model, which utilizes the interaction among users, time and users' continuous check-in behaviors to build functional relationship.

Firstly, we define an evaluation function $h(u, t, l_{cur}, l_{next})$ to express the degree of correlation among users, time, current POI location and the POI location that users may visit in the next moment. Where the value of function is the evaluation results of possibility that the user u is at POI's location l_{cur} at the time t and at POI's location l_{next} at the next moment. Here, $h: U \times T \times L \times L \rightarrow R$ can map a four tuple $(u, t, l_{cur}, l_{next})$ to a real value. The four tuple $(u, t, l_{cur}, l_{next})$ is a check-in for the user u at the POI's location l_{cur} at time t and at the POI's location l_{next} at time $t + 1$. Secondly, the interaction among users, time, current location and next location is used to describe the correlation effect of the function, and its evaluation function form is:

$$h(u, t, l_{cur}, l_{next}) = U_u^T \cdot L_{cur} + T_t^T \cdot L_{cur} + U_u^T \cdot T_t + U_u^T \cdot L_{next} + T_t^T \cdot L_{next} + L_{cur}^T \cdot L_{next} \tag{8}$$

Where $U_u, T_t, L_{cur}, L_{next} \in \mathbb{R}^k$ are respectively denote that the latent vectors of user u, time t, the current POI's location l_{cur}, the next POI's location l_{next}, and k represents the latent vector dimension. In addition, the inner product $U_u^T \cdot L_{cur}$ represents the interaction between the user and the current POI's location. $T_t^T L_{cur}$ represents the interaction between time segmentation and the current POI's location. The rest are similar to them. Since the first three items $U_u^T L_{cur}$, $T_t^T L_{cur}$, and $U_u^T T_t$ are independent of the next POI's location vector L_{next}, the calculation of the next location is not affected by their calculation results. Accordingly, in order to reduce the amount of calculation of the model, the evaluation function can be simplified. In addition, the influence of spatial information is only related to the interaction of the current POI's location and the next POI's location at successive locations, it is also independent of other interaction results. Therefore, the model can be further optimized by adding the modeling results of spatial effects. $\zeta_{lcur}(l)$ can be used as a weight value to adjust the model with user's successive check-in behaviors.

$$h(u, t, l_{cur}, l_{next}) = U_u^T \cdot L_{next} + T_t^T \cdot L_{next} + \zeta_{l_{cur}}(l) * L_{cur}^T \cdot L_{next} \tag{9}$$

3.5   Model Derivation

Inspired by the Bayesian Personalized Ranking [19], according to the personalized needs of different users, combined with the defined model, we design two methods based on partial order relationship to learn the model.

Optimization Objective Based on Locational Partial Order
Suppose that a user in the current POI location chooses the next POI location to be visited according to the following conditions: Users prefer to go to places they have been to rather than places they have never been to, or they prefer to go to places with high POI access frequency rather than places with low access frequency. Thus, the following conclusions can be obtained: for the h function, the score of h function for have visited POI is higher than that of a new POI, or the score of h function with high POI access frequency is larger than that with low POI access frequency.

Accordingly, we first define locational partial order. Assume that there is a partial order relationship $l_{nexti} \succ_{u,t,l_{cur}} l_{nextj}$. For $\forall l_{nexti}, l_{nextj} \in L$, user u at location l_{cur} at time t is always prefer to visit the next POI's location l_{nexti} that has been visited (or high POI access frequency) rather than the POI's location l_{nextj} that has not been visited (or low POI access frequency). According to the definition of the h function, there must be $h(u, t, l_{cur}, l_{nexti}) > h(u, t, l_{cur}, l_{nextj})$. We leverage the MAP method to calculate the parameters of the optimal model when we learn the model based on locational partial order.

$$O := \underset{\theta}{\arg\max} \, p(\theta | l_{nexti} \succ_{u,t,l_{cur}} l_{nextj}) \propto p(h(u, t, l_{cur}, l_{nexti}) > h(u, t, l_{cur}, l_{nextj}) | \theta) p(\theta) \tag{10}$$

$$p(h(u, t, l_{cur}, l_{nexti}) > h(u, t, l_{cur}, l_{nextj}) | \theta) = \frac{1}{1 + e^{-(h(u,t,l_{cur},l_{nexti}) - h(u,t,l_{cur},l_{nextj}))}} \tag{11}$$

Where logistic function $g(x) = 1/1 + e^{-x}$ to map it to $(0, 1)$, θ is the set of parameters, $p(\theta)$ is the prior probability, it satisfies the normal distribution $\theta \sim N(0, \sigma^2)$. Assume that for user u, each tuple $(u, t, l_{cur} \cdot l_{next})$ is independent, so the product of independent vectors can be used to further adjust and express the optimization objective of our model. Therefore, we use the multiple information such as User (abbr. as U), Temporal (abbr. as Te) and Spatial (abbr. as Sp) to establish a successive POI recommendation optimization target based on locational partial order, as shown in formula (12).

$$O_{UTeSp}^{L} := \arg\min_{\theta} \sum_{(u,t,l_{cur},l_{next}) \subset U \times T \times L \times L} - \ln g(h(u,t,l_{cur},l_{nexti}) - h(u,t,l_{cur},l_{nextj})) + \lambda \|\theta\|_F^2$$

$$(12)$$

Where λ is the coefficient of the regularization term associated with θ.

Optimization Objective Based on Temporal Partial Order
We also design a method to establish the objective function by using the temporal partial. If a user is at the current POI's location to decide where to go next. For the next POI that the user will visit, he always prefers the time period with high POI access frequency to travel instead of the time period with low access frequency. For example, users may be more willing to visit during the prime time of the mall, which is more in line by human travel habits. We find that the results of recommendation in different time periods are different and depend on the current location of users. The results of recommendation model can change dynamically over time.

According to TDP-HC algorithm, we define a temporal partial order. It is assumed that there is a time sequence relationship $t_i \succ_{u,l_{cur},l_{next}} t_j (\forall t_i, t_j \in T)$, t_i, t_j represent different time segments, and T is a set of time segments according to the optimal partition results. For user u, based on the current POI's location l_{cur}, he always preferentially selects l_{next} at time segment t_i with a high POI access frequency instead of time segment t_j with a low access frequency. When $\forall t_i, t_j \in T$, u can select the different time segment to visit the next POI's location. According to the definition of the function h, there must be $h(u, l_{cur}.l_{next}, t_i) > h(u, l_{cur}.l_{next}, t_j)$. Similar to the method of locational partial order, we still leverage the MAP method to calculate the parameters of the optimal model when we learn the model based on temporal partial order.

$$O := \arg\max_{\theta} p(\theta | t_i >_{u,l_{cur},l_{next}} t_j) \propto p(h(u, l_{cur}, l_{next}, t_i) > h(u, l_{cur}, l_{next}, t_j) | \theta) p(\theta) \quad (13)$$

Similarly, using the logistic function to map, the following results can be obtained.

$$p(h(u, l_{cur}, l_{next}, t_i) > h(u, l_{cur}, l_{next}, t_j) | \theta) = \frac{1}{1 + e^{-(h(u,l_{cur},l_{next},t_i) - h(u,l_{cur},l_{next},t_j))}} \quad (14)$$

Next, we further refine the optimize the objective function. Similar to the method based on locational partial order, integrating the influence of User, Temporal and Spatial information, a successive POI recommendation optimization objective O_{UTeSp}^{T} based on temporal partial order is established. Assume that the four tuple $(u, t, l_{cur}.l_{next})$ of all check-in records for each user u is independent, it has:

$$O^T_{UTeSp} := \arg\min_{\theta} \sum_{(u,l_{cur},l_{next},t) \subset U \times L \times L \times T} -\ln g(h(u,l_{cur},l_{next},t_i) - h(u,l_{cur},l_{next},t_j)) + \lambda\|\theta\|^2_F$$

(15)

3.6 Model Learning

In order to make the results of optimization more effectively, the stochastic gradient descent method can be used to learn the parameters in the UTeSp model. We learn the optimization objectives obtained based on locational partial order and temporal partial order respectively.

For each $di \in$ D, we update each parameter associated with the UTeSp model by following derivation.

$$\theta = \theta - \gamma\left(\frac{\partial}{\partial\theta}(\ln g(h(u,t,l_{cur},l_{nexti}) - h(u,t,l_{cur},l_{nextj})) + \lambda\|\theta\|^2_F)\right)$$

(16)

Where γ is the step size of stochastic gradient descent method, and the parameters $U_u, T_t, L_{cur}, L_{next} \in \theta$.

4 Experimental Results and Analysis

4.1 Data Description

In this paper, two real datasets Foursquare and Gowalla in LBSN are selected to verify the performance of UTeSp model. In the Gowalla dataset, we collect a total of 6,442,890 check-in records from February 2009 to October 2010, in which the friend relationship included 196,491 nodes and 950,327 edges. The Foursquare dataset is also a real dataset based on LBSN.

The data selected in this paper is derived from some experimental data of Gao et al. [20]. In order to obtain better model performance, we remove the data that the number of user check-ins is less than 15 times and the number of check-ins on POI is less than 15 times on two datasets. And the statistics after processing are shown in Table 2. The whole dataset is divided into three parts: training set 70%, verification set 10%, testing set 20%.

Table 2. Statistics of Foursquare and Gowalla

Category	Foursquare	Gowalla
Number of users	12217	16287
Number of POIs	20592	21640
Number of check-in records	1260809	2104464

4.2   Evaluation Criteria

The UTeSp model selects the F1-score to evaluate the performance of model, which is a weighted average of the precision and recall. When the value of the F1-score is high, the experimental method is more effective.

$$F1 = \frac{2 * Precision * Recall}{Precision + Recall} \tag{17}$$

4.3   Other Models of the Current Mainstream

In order to verify the validity of the model, the UTeSp model selects the following three current mainstream successive POI recommendation models for comparison. FPMC [5] proposes a personalized successive POI recommendation model that combines Markov chain and matrix factorization methods. PRME [8] is the first successive POI recommendation model based on metric learning, which combines sequence transformation with user preference and geographical influence. STELLAR [9] is a time-space successive POI recommendation.

4.4   Experimental Parameter Settings

Our UTeSp model uses two different methods of partial order for model learning. In order to better distinguish locational partial order method and the temporal partial order method, we named them UTeSp_S and UTeSp_T, respectively.

In the two datasets, a large number of experiments were performed on our two models to optimize. The grid search method is used to find the best feature dimension and regularization coefficient a for the two models. For dimension K, it increments from 20 to 200 dimensions by 20 dimensions to find the best feature dimension. Meanwhile, we find the best regularization coefficient λ from 0.0001 to 0.1, increasing by 5 times. The following different parameters can be set to obtain the optimal performance. The parameters for the UTeSp_S model are set as follows: $K = 80$, $\lambda = 0.001$, $\gamma = 0.003$. Similarly, the parameters for the UTeSp_T model are set as follows: $K = 60$, $\lambda = 0.001$, $\gamma = 0.005$. For the convenience of calculation, we take $\beta = 1/2$.

4.5   Experimental Analysis

Performance Comparison with Other Models

In the experiment, the performance of UTeSp_S and UTeSp_T are compared with the current mainstream successive POI recommendation models on the F1-score of topN. Figure 1 shows the comparison results.

The UTeSp_S and the UTeSp_T have the best performance of all the models, STELLAR is the best model participating in the comparison, and FPMC is the weakest. In the Gowalla, compared with the earliest FPMC model, the performance of our UTeSp_S and STELLAR increased by 98.7% and 49.3% at F1@5 respectively, 0.9

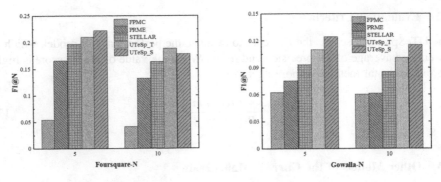

Fig. 1. The performance comparison of UTeSp and other models

times and 0.4 times on F1@10. In the Foursquare, the UTeSp_S and the UTeSp_T increased by about 3.0 times and 2.6 times compared to FPMC at F1@5, and by 4.3 times and 2.9 times respectively at F1@10. And, the performance of UTeSp_S model is 33.12% higher than STELLAR on the Gowalla at F1@5, the performance is improved by 12.33% on the Foursquare at F1@5.

Locational Partial Order Method and Temporal Partial Order Method

Figure 2 shows the comparison result between UTeSp_S and UTeSp_T. The experiment finds that UTeSp_S has 5.41% performance improvement compared with UTeSp_T at F1@5 in Foursquare. At F1@10, The former improved by 4.91% over the latter. In Gowalla, UTeSp_S is 12.96% higher than that of UTeSp_T at F1@5, and the former is almost 14.46% higher than the latter at F1@10.

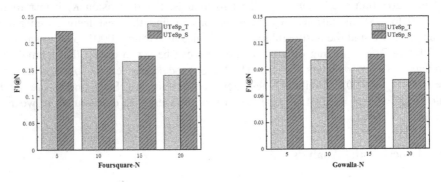

Fig. 2. The Performance comparison of UTeSp_S and UTeSp_T

In general, the method based on the locational partial order obtains more effective experimental results. Although UTeSp_T is slightly worse than UTeSp_S, it still obtains better performance than other mainstream models. Meanwhile, the method based on temporal partial order has better interpretability and can meet the different individual needs of users, so it is also considered as a good choice. After that, we use the better UTeSp_S model to replace the two models for analysis.

The Effect of User's Interest Preference on Recommendation Performance

Figure 3 shows the performance comparison of the UTeSp model with different user interest preferences. It has two forms in the UTeSp_S model. One is the UTeSp_S-normal model, which only uses the user's own internal interest preference. The other is our UTeSp_S, which uses the user's internal interest preference together with the influence of user's friends.

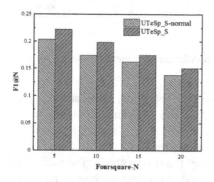 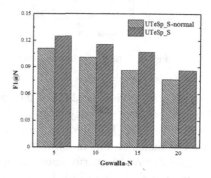

Fig. 3. The performance comparison of different user characteristics

After experimental analysis, we found that in Foursquare, UTeSp_S is 9.29% higher than UTeSp_S-normal at F1@5, and 14.01% at F1@10. In Gowalla, the former increased by 12.25% at F1@5 and by 14.34% at F1@10. It shows that both the user's internal interest preferences and the influence of user's friends can play a positive role in UTeSp_S. At the same time, the introduction of attention mechanism can better obtain the influence of friend users on target users, thus further accelerating the improvement of model performance.

The Effect of Different Time Partition Methods

Table 3 shows the results of static time partition and dynamic time partition. The former is a static manual partition method, while the latter is a combination of dynamic and manual partition.

Table 3. Statistical results of different time partition methods in Foursquare ($ns = 4 * 2$)

Event category	Regular time partition	Dynamic time partition
T1- Weekdays	00:00–06:00	00:09:23–05:23:45
T2- Weekdays	06:00–12:00	05:23:45–10:53:32
T3- Weekdays	12:00–18:00	10:53:32–17:54:06
T4- Weekdays	18:00–00:00	17:54:06–00:09:23
T5- Weekends	00:00–06:00	22:00:15–03:24:36
T6- Weekends	06:00–12:00	03:24:36–08:43:56
T7- Weekends	12:00–18:00	08:43:56–14:30:51
T8- Weekends	18:00–00:00	14:30:51–22:00:15

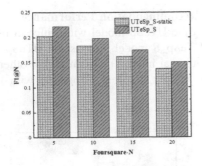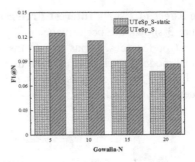

Fig. 4. The performance comparison on different time partition methods

Initially, we divide the 24 h of the day into four segments. Next, according to the difference between weekdays (Monday to Friday) and weekends (Saturday and Sunday), we use TDP_HC algorithm to divide the time segment into eight ($ns = 4 * 2$). There are similar results in Gowalla, we will not repeat them.

Figure 4 shows the performance comparison of the two methods. The method using static time partition is recorded as UTeSp_S-static, while the method using Dynamic time partition is our UTeSp_S model. The result shows that our dynamic time partition method can significantly improve the performance of the model, thus proving the effectiveness of our method.

The Effect of Spatial Information on Recommendation Performance

In spatial modeling, we design a $\zeta_{lcur}(l)$ to adjust the relationship between the user's current location and the next location to be visited. Figure 5 is the comparison result of whether the spatial information is applied to the UTeSp_S model. UTeSp_S and UTe_S represent the models with or without considering the influence of spatial information.

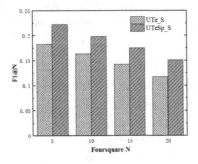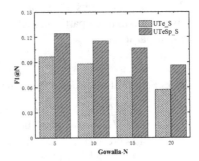

Fig. 5. The performance comparison by utilizing spatial influence

Compared to UTe_S model in Foursquare, UTeSp_S is 21.21% and 20.95% higher than F1@5 and F1@10. In Gowalla, UTeSp_S is also increased by 28.19% and 30.62% compared to UTe_S. The experimental results demonstrate that the spatial information is important in our model, and users' willingness to visit successive locations may be closely related to spatial distance. The influence of spatial information not only leads to performance improvements, but also conforms to the real life of the user's need.

5 Conclusions

In this paper, a dynamic UTeSp model based on user-level attention is proposed to solve the successive POI recommendation problem. In order to alleviate the data sparsity, we use the collaborative influence of multiple interactions to construct our model. At the same time, a new user interest preference and an effective TDP_HC algorithm is designed to achieve personalized user needs. In addition, we propose two interpretable personalized model learning methods to learn the model more effectively. The experimental results show that: (1) UTeSp model is obviously better than other mainstream successive recommendation models in performance. (2) The performance of successive POI recommendation model can be improved by the interaction of multiple information. In the future, we plan to use the effective information from multiple views such as textual content, visual images and the diversity of items to further study the successive POI recommendation problem in the graph neural network.

References

1. Bin, C., Gu, T., Sun, Y., Chang, L.: A personalized POI route recommendation system based on heterogeneous tourism data and sequential pattern mining. Multimedia Tools Appl. **78** (24), 35135–35156 (2019). https://doi.org/10.1007/s11042-019-08096-w
2. Guo, J., Zhang, W., Fan, W., et al.: Combining geographical and social influences with deep learning for personalized point-of-interest recommendation. J. Manag. Inf. Syst. **35**(4), 1121–1153 (2018)
3. Lu, Y., Huang, J.: GLR.: a graph-based latent representation model for successive POI recommendation. Future Gener. Comput. Syst. **102**, 230–244 (2020)
4. Lu, Y.S., Shih, W.Y., Gau, H.Y., et al.: On successive point-of-interest recommendation. World Wide **22**, 1151–1173 (2019)
5. Rendle, S., Freudenthaler, C., et al.: Factorizing personalized Markov chains for next-basket recommendation. In: The Web Conference 2010, pp. 811–820, Raleigh (2010)
6. Cheng, C., et al.: Where you like to go next: successive point-of-interest recommendation. In: Twenty-Third International Joint Conference on Artificial Intelligence 2013, pp. 2605–2611, Beijing (2013)
7. Liu, X., Liu, Y., Aberer, K., et al.: Personalized point-of-interest recommendation by mining users' preference transition. In: Conference on Information and Knowledge Management 2013, pp. 733–738. Springer, San Francisco (2013)
8. Feng, S., Li, X., Zeng, Y., et al.: Personalized ranking metric embedding for next new POI recommendation. In: International Conference on Artificial Intelligence 2015, pp. 2069–2075, Buenos Aires (2015)
9. Zhao, S., Zhao, T., Yang, H., et al.: STELLAR: spatial-temporal latent ranking for successive point-of-interest recommendation. In: National Conference on Artificial Intelligence 2016, pp. 315–321, Phoenix (2016)

10. He, J., Li, X., Liao, L., et al.: Next point-of-interest recommendation via a category-aware Listwise Bayesian Personalized Ranking. J. Comput. Sci. **28**, 206–216 (2018)
11. Wu, L., Sun, P., et al.: SocialGCN: an efficient graph convolutional network based model for social recommendation. https://arxiv.org/abs/1811.02815v2 July 2019
12. Wang, M., Zheng, X., et al.: Collaborative filtering with social exposure: a modular approach to social recommendation. https://arxiv.org/abs/1711.11458, November 2017
13. Rahmani, H.A., Aliannejadi, M., Ahmadian, S., et al.: LGLMF: Local Geographical based Logistic Matrix Factorization Model for POI Recommendation. https://arxiv.org/abs/1909.06667, September 2019
14. Zhang, Z., Liu, Y., Zhang, Z., et al.: Fused matrix factorization with multi-tag, social and geographical influences for POI recommendation. World Wide Web. **22**, 1135–1150 (2019)
15. Chen, J., Zhang, H., et al.: Attentive collaborative filtering: multimedia recommendation with item-and component-level attention. In: ACM SIGIR FORUM 2017, vol. 51, pp. 335–344. Springer, Tokyo (2017)
16. Huang, H., Dong, Y., Tang, J., et al.: Will triadic closure strengthen ties in social networks. ACM Trans. Knowl. Discov. Data **12**(3), 1–25 (2018)
17. Guo, L., Wen, Y., Liu, F., et al.: Location perspective-based neighborhood-aware POI recommendation in location-based social networks. Soft. Comput. **23**, 11935–11945 (2019)
18. Yuan, Q., Cong, G., Ma, Z., et al.: Time-aware point-of-interest recommendation. In: International ACM SIGIR Conference on Research and Development in Information Retrieval 2013, pp. 363–372. Springer, Dublin (2013)
19. Li, H., Diao, X., Cao, J., et al.: Tag-aware recommendation based on Bayesian personalized ranking and feature mapping. Intell. Data Anal. **23**(3), 641–659 (2019)
20. Gao, H., Tang, J., Liu, H., et al.: gSCorr: modeling geo-social correlations for new check-ins on location-based social networks. In: Conference on Information and Knowledge Management 2012, pp. 1582–1586. Springer, Maui (2012)

Application Track

Chemical XAI to Discover Probable Compounds' Spaces Based on Mixture of Multiple Mutated Exemplars and Bioassay Existence Ratio

Takashi Isobe[1,2,3]([✉]) and Yoshihiro Okada[2]

[1] Hitachi High-Tech America, Inc., Pleasanton, CA 95488, USA
takashi.isobe.sw@hitachi-hightech.com
[2] Hitachi High-Tech Solutions Corporation, Chuo-ku, Tokyo 104-6031, Japan
[3] Hitachi High-Tech Corporation, Minato-ku, Tokyo 105-8717, Japan

Abstract. Chemical industry pays much cost and long time to develop a new compound having aimed biological activity. On average, 10,000 candidates are prepared for each successful compound. The developers need to efficiently discover initial candidates before actual synthesis, optimization and evaluation. We developed a similarity-based chemical XAI system to discover probable compounds' spaces based on mixture of multiple mutated exemplars and bioassay existence ratio. Our system piles up 4.6k exemplars and 100M public DB compounds into vectors including 41 features. Users input two biologically active sets of exemplars customized with differentiated features. Our XAI extracts compounds' spaces simultaneously similar to multiple customized exemplars using vectors' distances and predicts their biological activity and target with the probability shown as existence ratio of bioassay that is the information of biological activity and target obtained from public DB or literature including related specific text string. The basis of prediction is explainable by showing biological activity and target of similar compounds included in the extracted spaces. The mixture of multiple mutated exemplars and bioassay existence ratio shown as probability with the basis of prediction can help the developers extract probable compounds' spaces having biological activity from unknown space. The response time to extract the spaces between two sets of 128 exemplars and 100M public DB compounds was 9 min using single GPU with HDD read and 1.5 min on memory. The bioassay existence ratio of extracted spaces was 2–14 times higher than the average of public ones. The correlation coefficient and R2 between predicted and actual pIC50 of biological activity were 0.85 and 0.73 using randomly selected 64 compounds. Our XAI discovered probable compounds' spaces from large space at high speed and probability.

Keywords: AI · Explainable AI · XAI · Compound · Discovery · Virtual Screening

© Springer Nature Switzerland AG 2020
S. Nepal et al. (Eds.): BIGDATA 2020, LNCS 12402, pp. 177–189, 2020.
https://doi.org/10.1007/978-3-030-59612-5_13

1 Background and Purpose

Chemical industry pays much cost and long time to develop a new compound having aimed biological activity. The developers specify the target and discover lead compounds as candidates. After preparing them, they actually synthesize, optimize and evaluate them. On average, 10,000 candidates are prepared and tested for each successful compound [5, 6]. The developers need to efficiently discover lead compounds becoming initial candidates before actual synthesis, optimization and evaluation.

Chemical industry uses VS (Virtual Screening) [1] for discovery of new lead compounds having aimed biological activity. Virtual screening is classified into two techniques based on similarity to exemplars [8, 14, 16] and docking with target protein [10]. Both techniques use structure, polarity, physical and chemical properties of chemical compounds.

The similarity-based technique can discover probable candidates at high accuracy in well-known compounds' space having much data about biological activity while it has difficulty in unknown compounds' space where the developers have no information about biological activity that is often closed by each organization.

The docking-based technique can discover candidates having new structure in unknown compounds' space without any information about biological activity while it needs the detailed information of target proteins and much calculation resource according to rotatable ligands and deformable proteins.

We aimed to develop a similarity-based chemical XAI (Explainable AI) system to cost-effectively discover probable compounds' space having new structure from larger unknown compounds' space at higher speed and probability based on mixture of multiple mutated exemplars and bioassay existence ratio shown as probability with the basis of prediction.

2 Related Works

Virtual screening is classified into 2D and 3D techniques in addition to similarity and docking. 2D-based technique includes LBVS (Ligand-Based Virtual Screening) [12] and CGBVS (Chemical Genomics-Based Virtual Screening) [9]. 3D-based technique includes SBVS (Structure-Based Virtual Screening) [10] and PBVS (Pharmacophore-Based Virtual Screening) [4, 11].

LBVS including QSAR (Quantitative Structure-Activity Relationship) [17] discovers probable candidates based on similarity to existing compounds whose biological activity are known using numeric array transferred from the features of chemical compounds. The technique of fingerprint is often used as the numeric array [3]. The fingerprint-based technique shows the existence of each structural feature as bit array of 1 or 0. Even if there is a great difference between two compounds in the number of polarity like oxo group, similarity becomes same.

CGBVS recognizes the pattern of biological activity and target by each combination of ligand and target protein using matrixes arrayed from structures and features of not only ligand but also target protein.

SBVS predicts the intensity of biological activity by simulating structural binding-conformation of ligands to binding-pocket of target protein.

PBVS creates the model of 3D pharmacophore based on polarity, hydrogen bond and Van der Waals force between existing ligands and target protein whose biological activity and target are known. It discovers probable candidates having similar features using that model.

LBVS bases similarity to exemplars or ligands and needs no information about targets. Therefore, even if the developers have no detailed information about targets, it can discover candidates. On the other hand, it has difficulties in unknown compounds' space having new structure where the developers have no information about aimed biological activity.

CGBVS, SBVS and PBVS recognize the conformation between exemplars/ligands and target protein using the information of structure and polarity. They can predict the biological activity of new structural compounds in unknown compounds' space where the developers have no information about aimed biological activity. On the other hand, they are not available without the detailed and structural information of target protein. The developers often have no detailed and structural information about target proteins, enzymes or biosynthetic pathway. Therefore, applicable cases are limited.

SBVS and PBVS avoid the exponential increase of calculation by assuming deformable protein as rigid body, limiting the combinations of rotatable ligands or eliminating the hydrogen bond of water molecules. Therefore, the calculation becomes approximate and the accuracy often becomes low. PBVS considers not only structural binding-conformation but also binding-energy of hydrogen bond and Van der Waals force. Therefore, PBVS needs much more calculation than SBVS.

A paper about the pitfalls of VS [1] reports that 2D-based technique is often better than 3D-based one. It also shows that similarity-based technique often performs better than docking-based one. In the other paper [2], ligand or exemplar-based technique shows the similar accuracy as 3D docking-based ones.

3 Proposed Chemical XAI System

3.1 System Configuration

The following Fig. 1 shows the configuration of our chemical XAI system to discover probable compounds' spaces based on mixture of multiple mutated exemplars and probability shown as bioassay existence ratio with the basis of prediction. Our system works as a cloud service of SaaS (Software as a Service) with other services such as NaaS (Network as a Service) [18]. It piles up 4.6K exemplars and 100M public DB compounds into feature vectors including 41 features. Users input two biologically active sets of exemplars customized with differentiated structure or identifier into XAI application of cloud, and after then, they receive the output from the cloud.

Our XAI extracts compounds' spaces simultaneously similar to multiple customized exemplars based on Euclidean vectors' distances tuning weight in some features. In addition, it predicts biological activity intensity (IC50: half maximal inhibitory concentration) and target with probability by each extracted compounds' space based on K-NN (Nearest Neighbor). The probability is shown as existence ratio of bioassay that is the information of biological activity and target obtained from public DB or literature including related specific text string.

The basis of prediction is explainable by showing biological activity and target of similar compounds included in the extracted spaces. The mixture of multiple mutated exemplars and bioassay existence ratio shown as probability with the basis of prediction can help the developers extract probable compounds' spaces having biological activity from unknown compounds' space.

Our system comprises of feature extractor, public DB table, exemplar table and XAI application. Feature extractor generates feature vectors including the features of structure, pharmacophore, polarity, physical and chemical properties from SDF files recording compound data of public DB [7] and stores them to public DB table. Public DB table has the information of biological activity and target by each compound's feature vector. Exemplar table records the combinations of compound's feature vector, biological activity and target obtained from specific field DB of biologically active compounds such as drug, food, inhibitor and other.

XAI application extracts probable compounds' spaces based on similarity calculated with brute force by feature vectors' distances between customized exemplars and public DB compounds. Similarity considers the easiness of docking with target using feature vectors including pharmacophore and polarity. The extracted compounds' spaces have the features mixed between multiple exemplars whose part can be mutated by customization. After the extraction, our XAI predicts the biological activity intensity (IC50) and target with the probability shown as bioassay existence ratio by each extracted compounds' space based on biological activity and target of similar compounds included in the extracted space while showing the similar compounds with the biological activity and target as the basis of prediction.

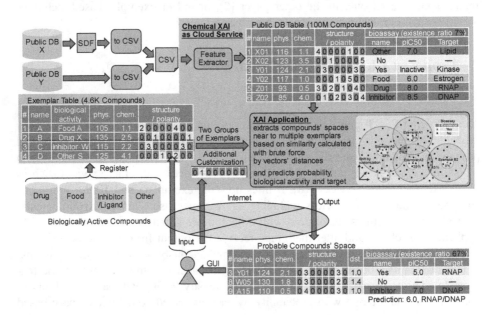

Fig. 1. System configuration

Explainability of causality between customized exemplars and probability of extracted compounds' spaces is very important to suggest the direction about how the users customize or mix exemplars for the next trial. Our XAI can explain the basis of the probability and prediction by showing bioassay existence, biological activity and target of similar compounds included in the extracted compounds' spaces. Moreover, the users have to repeat the discovery of highly probable unknown compounds' spaces by gradually changing the mixture of exemplars partially differentiated with customization and adjusting the position of the extracted compounds' spaces based on their experience and intuition. The mixture of multiple mutated exemplars and probability shown as bioassay existence ratio with the basis of prediction can help the developers extract new probable compounds' spaces having biological activity from unknown compounds' space where they have no biological activity data.

3.2 Feature Vectors with Biological Activity and Target

Our system stores feature vectors of compounds with bioassay including biological activity and target into public DB and exemplar tables.

Public DB table stores 100M feature vectors generated from SDF (Structure Data File) including compound data of public DB [7]. Feature vector comprises of 41 features considering Lipinski's rule of five [19]. The vector includes the number of ring structures classified by the number of elements, nitrogen, oxygen and sulfur, the number of chain structures classified by substituent, functional, characteristic and joint group, the number of pharmacophore combining various structures and polarities, and physical/chemical properties such as molar mass, boiling point, freezing point, vapor pressure, density, water solubility, organic solvent solubility, thermal stability, acid alkalinity and spectral by each chemical compound. The vector has the features considering the easiness of docking with target such as hydrogen bond, Van der Waals force and hooking structure.

Each feature vector has the bioassay that is the information of biological activity, its intensity and target obtained from not only public DB but also literature such as patents and papers including related specific text string. Biological activity shows the kind of biological effect. The intensity of biological activity uses the value of IC50 (half maximal inhibitory concentration). Target includes protein, enzyme and biosynthetic pathway related to biosynthesis or proteolysis of protein or lipid in the biological field such as mitochondria, chloroplast, nucleus, ribosome, lysosome and phosphate cycle by each biological species. The prediction of probability, biological activity and target can increase the accuracy using not only public data but also users' own data.

Exemplar table stores 4.6K combinations of compounds' feature vector, biological activity and target obtained from public DB of specific fields related to biological activities such as drug, food, inhibitor and other which the users especially focus on.

Table 1 shows the number of compounds registered to public DB and exemplar tables. Bioassay existence means the number of compounds that are known to have biological activities such as drug, food additive, inhibitor or others. Intensity of biological activity (IC50) means the number of compounds having the information of the intensity of biological activities. Target means the number of compounds having the information of target. Even if the users have no information about biological activities and target, our XAI can discover probable compounds' spaces by predicting their probability, biological activity and target using the information obtained from various public DB.

Table 1. Proportion of each data in public DB and exemplar table

Data	Number of compounds
Public DB table	99,914,173
Bioassay existence	6,993,164 (7.0%)
Intensity of biological activity (IC50)	2,439,289 (2.4%)
Target	2,834,792 (2.8%)
Exemplar table	4,612

3.3 Extracted Compounds' Spaces and Bioassay Existence Ratio

Our XAI application extracts four kinds of compounds' spaces including similar vectors by calculating Euclidean vector distances with brute force between two biologically active sets of exemplar and public DB vectors. Weight of each feature in vectors can be tuned by user-specified values. First space is near to single biologically active group of exemplar vectors. Second one includes typical features simultaneously near to multiple vectors included in single biologically active group of exemplar vectors. Third one is mixed space simultaneously near to two biologically active groups of exemplar vectors. Fourth one is mutated space designed and shifted with additional customization. It may have the features mixed between multiple exemplars whose part is mutated by customization (See Fig. 2).

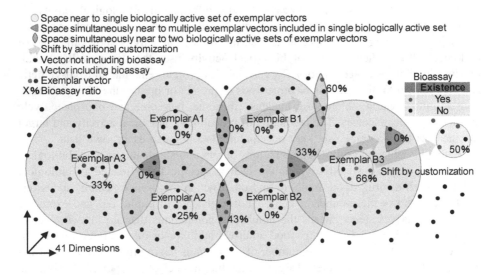

Fig. 2. Extracted compounds' spaces and bioassay existence ratio

After the extraction, our XAI application calculates the bioassay existence ratio using the bioassay of similar compounds included in each extracted space and decides the probability based on the deviation of the bioassay existence ratio of each extracted space compared to the average of entire space.

Similarity-based technique generally performs well in well-known spaces where users have much information about aimed biological activity. On the other hand, it has difficulties predicting probability in unknown spaces having new structure where users have no information about aimed biological activity. The information of aimed biological activity is often closed by each organization while the data of other biological activity is often opened. Even if there is no data of aimed biological activity around exemplars, we can approximately estimate the probability using bioassay existence ratio considering various biological activities. Our XAI covers the disadvantage of similarity-based technique by predicting probability using bioassay existence ratio considering various biological activity.

In some cases, users may not have any information of aimed or other biological activities around exemplars. Even in such cases, our XAI can explore the mixed or mutated compounds' spaces simultaneously near to two biologically active groups of exemplars partially customized with differentiated features and can help the developers discover probable compounds' space having new structure from unknown compounds' space far away from exemplars by predicting the biological activity with probability shown as bioassay existence ratio.

3.4 Prediction of Biological Activity and Target

Our XAI predicts the biological activity intensity ($pIC50 = -\log_{10} IC50$) and target with the probability shown as bioassay existence ratio by each extracted compounds' space based on biological activity and target of similar compounds included in the extracted space while showing similar compounds with biological activity and target as the basis of prediction.

Prediction of biological activity intensity pIC50 uses the same K-NN technique as rehabilitation AI [13] that is our other AI system. Our XAI uses similar compounds having IC50 data from similar compounds included in extracted compounds' space. It classifies the extracted similar compounds into three patterns according to the value of IC50 and calculates the predicted value of pIC50 using the similar compounds included in the pattern deviating significantly from the average proportion of public DB table's compounds.

Prediction of target also uses similar compounds having target data from similar compounds included in extracted compounds' space. Our XAI summarizes the total number by each kind of target and shows the targets occupying large proportion as predicted target.

3.5 Importance of Explainability for Mixture of Mutated Exemplars

Explainability of causality between customized exemplars and probability of extracted spaces is very important to suggest the direction about how the users customize or mix exemplar for the next trial. Our XAI can explain the basis of the probability and prediction by showing bioassay existence, biological activity and target of similar compounds included in the extracted compounds' spaces.

From biological activity and target of similar compounds shown as the basis of prediction, the users can acquire the direction about how to customize or mix exemplars for the next trial and repeat the discovery of highly probable unknown compounds' spaces by gradually changing the mixture of exemplars shifted with partial customization and adjusting the position of mixed or mutated compounds' spaces based on their experience and intuition. The process of discovery can be automated by automatically assigning mixture or mutation of exemplars with brute force.

The mixture of multiple mutated exemplars and probability shown as bioassay existence ratio with the basis of prediction can help the developers extract new probable compounds' spaces having biological activity from unknown compounds' space where they have no biological activity data. The method of repeating the discovery of probable compounds' spaces by mixing or mutating feature vectors between multiple exemplars is attempting to recreate the way how living creatures have acquired survival advantages in long-term evolution by repeating mixture of male and female DNAs (Deoxyribonucleic Acid) or mutation of them. It is also similar to humans' intuition or creativity generating new concepts by mixing multiple various past memories in their brains.

4   Evaluation

4.1   Execution Time

Execution time is important to improve efficiency of developing a new compound because developers repeat the discovery of highly probable unknown compounds' spaces by gradually tuning the mixture or customization of exemplars. We measured the response time to extract the spaces including similar vectors between two biologically active groups of 128 exemplars and 100M public DB compounds. The response time was 9 min using single GPU (Tesla P100) with data read from HDD and 1.5 min on memory. Our XAI system could screen compounds at 55–333 times faster speed per single GPU from 16 times larger space compared to the result of 15 min for 6M compounds over dual-GPU computer reported in a past exemplar/ligand-based paper [2] that supported the prediction of probability. We also achieved 7.5–45 times faster speed per single GPU compared to the result of 18 min for 1330M compounds over 50 GPUs reported in the other paper [15] that didn't support the prediction of probability, target or biological activity.

4.2   Prediction of Probability

We also evaluated the bioassay existence ratio by each compounds' space extracted using mixtures of drug, food additive, inhibitor and other designed with additional customization as exemplars. The results are shown in Figs. 3, 4, 5 and 6.

The space near to single exemplar had 2–6 times higher bioassay existence ratio than the average of public DB compounds' space where smaller CID tended to be older

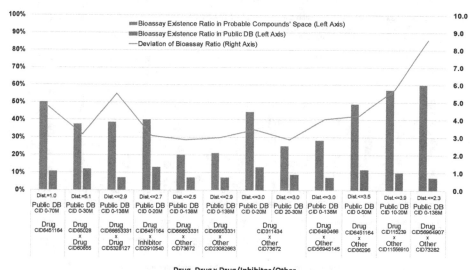

Fig. 3. Bioassay existence ratio and its deviation in spaces using drug as first exemplar

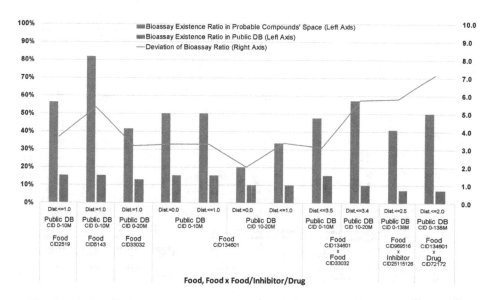

Fig. 4. Bioassay existence ratio and its deviation in spaces using food as first exemplar

and have higher bioassay existence ratio. The overlapped space simultaneously near to two biologically active groups of exemplars had 2–14 times higher bioassay existence ratio than the average of public DB compounds' space. In the space near to unknown structural exemplar compound designed by adding some groups to existing compounds, the bioassay existence ratio was 4–7 times higher.

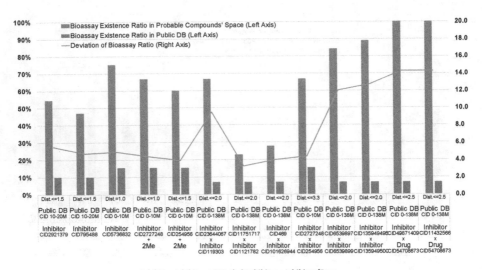

Fig. 5. Bioassay existence ratio and its deviation in spaces using inhibitor as first exemplar

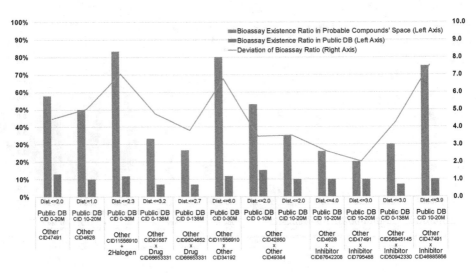

Fig. 6. Bioassay existence ratio and its deviation in spaces using other as first exemplar

A past paper comparing exemplar/ligand-based and docking-based virtual screening [2] used the ratio of true positive ratio (TPR) and false positive ratio (FPR) as probability. That ratio was 2–10 times in the area where FPR was 5–15%. On the other hand, our XAI can't acquire the ratio of TPR and FPR because it aims at discovering probable compounds' spaces from unknown compounds' space where users have no information of aimed biological activity. Therefore, our XAI uses the deviation of bioassay existence ratio of extracted compounds' space against the average of public DB compounds' space as the probability instead of the ratio of TPR and FPR. That deviation was 2–14 times similar to the past exemplar-based and docking-based virtual screening. Our XAI could discover probable compounds' space at the similar probability to the past techniques from unknown compounds' space where users had no information of biological activity.

4.3 Prediction of Biological Activity

In the probable compounds' spaces, our XAI can predict the intensity of biological activity called as pIC50 = $-\log_{10}$ IC50 (half maximal inhibitory concentration) using IC50 value of similar different compounds. We confirmed that the correlation coefficient and R2 between predicted and actual pIC50 values were 0.85 and 0.73 using randomly selected 64 compounds having actual IC50 value (See Fig. 7).

The average R2 of 0.47 is reported in a past paper [17]. Our XAI could predict the intensity of biological activity at high accuracy in unknown compounds' space where users had no information of biological activity.

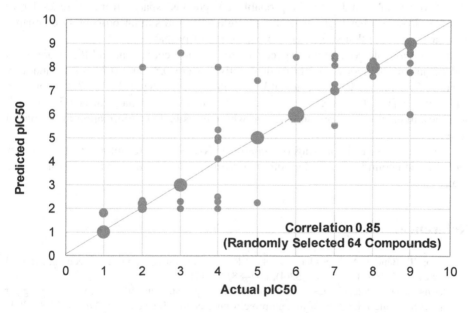

Fig. 7. Correlation between actual and predicted pIC50

5 Conclusion

We developed a similarity-based chemical XAI system to discover probable compounds' spaces based on mixture of multiple mutated exemplars and bioassay existence ratio.

188   T. Isobe and Y. Okada

Our XAI system works as a cloud service. It piles up 4.6K exemplars and 100M public DB compounds into vectors including 41 features. Users input two biologically active sets of exemplars customized with differentiated features into XAI application of cloud, and after then, they receive the output from the application.

Our XAI application extracts compounds' spaces simultaneously similar to multiple customized exemplars using vectors' distances and predicts their biological activity and target with the probability shown as existence ratio of bioassay that is the information of biological activity and target obtained from public DB or literature including related specific text string.

The basis of prediction is explainable by showing biological activity and target of similar compounds included in the extracted spaces. The mixture of multiple mutated exemplars and bioassay existence ratio shown as probability with the basis of prediction can help the developers extract probable spaces having biological activity from unknown compounds' space.

The response time to extract the compounds' spaces between two sets of 128 exemplars and 100M public DB compounds was 9 min using single GPU with HDD read and 1.5 min on memory. The bioassay existence ratio of extracted compounds' spaces was 2–14 times higher than the average of public DB compounds' space. Our XAI system could discover probable compounds' spaces from 16 times larger space at 55–333 times higher speed per single GPU and similar probability compared to a past paper [2] that supported the prediction of probability.

In the probable compounds' spaces, the correlation coefficient and R2 of predicted and actual pIC50 intensity of biological activity were 0.85 and 0.73 using randomly selected 64 compounds having actual IC50 value. The average R2 of 0.47 is reported in a past paper [17]. Our XAI could predict the intensity of biological activity at high accuracy in unknown compounds' space where users had no information of biological activity.

We are currently improving our chemical XAI system to automatically discover probable compounds' spaces by automating the mixture of mutated exemplars as the future task.

References

1. Scior, T., Bender, A., Tresadern, G., et al.: Recognizing pitfalls in virtual screening: a critical review. J. Chem. Inf. Model. **52**(4), 867–881 (2012)
2. Johnson, D.K., Karanicolas, J.: Ultra-high-throughput structure-based virtual screening for small-molecule inhibitors of protein-protein interactions. J. Chem. Inf. Model. **56**(2), 399–411 (2016)
3. Willett, P.: Similarity-based virtual screening using 2D fingerprints. In: Drug Discovery Today, vol. 11, no. 23–24, pp. 1046–1053. Elsevier (2006)
4. Matter, H., Poetter, T.: Comparing 3D pharmacophore triplets and 2D fingerprints for selecting diverse compound subsets. J. Chem. Inf. Comput. Sci. **39**(6), 1211–1225 (1999)
5. Hambley, T.W.: The influence of structure on the activity and toxicity of Pt anti-cancer drugs. In: Coordination Chemistry Reviews, vol. 166, pp. 181–223. Elsevier (1997)
6. Wolff, M., McPherson, A.: Antibody-directed drug discovery. Nature **345**, 365–366 (1990)
7. PubChem Homepage. https://pubchem.ncbi.nlm.nih.gov/. Accessed 03 Dec 2019

8. Willett, P., Barnard, J.M., Downs, G.M.: Chemical similarity searching. J. Chem. Inf. Comput. Sci. **38**(6), 983–996 (1998)
9. Hamanaka, M., et al.: CGBVS-DNN: prediction of compound-protein interactions based on deep learning. Mol. Inf. **36**(1–2), 1600045 (2017)
10. Elokely, K.M., Doerksen, R.J.: Docking challenge: protein sampling and molecular docking performance. J. Chem. Inf. Model. **53**(8), 1934–1945 (2013)
11. Wang, J., et al.: Pharmacophore-based virtual screening and biological evaluation of small molecule inhibitors for protein arginine methylation. J. Med. Chem. **55**(18), 7978–7987 (2012)
12. Kurczyk, A., Warszycki, D., Musiol, R., Kafel, R., Bojarski, A.J., Polanski, J.: Ligand-based virtual screening in a search for novel anti-HIV-1 chemotypes. J. Chem. Inf. Model. **55**(10), 2168–2177 (2015)
13. Isobe, T., Okada, Y.: Medical AI system to assist rehabilitation therapy. In: Perner, P. (ed.) ICDM 2018. LNCS (LNAI), vol. 10933, pp. 266–271. Springer, Cham (2018). https://doi.org/10.1007/978-3-319-95786-9_20
14. Lo, Y.-C., Rensi, S.E., Torng, W., Altman, R.B.: Machine learning in chemoinformatics and drug discovery. Drug Discov. Today **23**(8), 1538–1546 (2018)
15. Grebner, C., Malmerberg, E., Shewmaker, A., Batista, J., Nicholls, A., Sadowski, J.: Virtual screening in the cloud: how big is big enough. J. Chem. Inf. Model. (2019)
16. Kristensen, T.G., Nielsen, J., Pedersen, C.N.S.: Methods for similarity-based virtual screening. Comput. Struct. Biotechnol. J. **5**(6), e201302009 (2013)
17. Kato, Y., Hamada, S., Goto, H.: Validation study of QSAR/DNN models using the competition datasets. Mol. Inf. **2019**, 30 (2019)
18. Isobe, T., Tanida, N., Oishi, Y., Yoshida, K.: TCP acceleration technology for cloud computing: Algorithm, performance evaluation in real network. In: 2014 International Conference on Advanced Technologies for Communications (ATC 2014), pp. 714–719. IEEE (2014)
19. Lipinski, C.A., Lombardo, F., Dominy, B.W., Feeney, P.J.: Experimental and computational approaches to estimate solubility and permeability in drug discovery and development settings. Adv. Drug Deliv. Rev. **23**(1–3), 3–25 (1997)

Clinical Trials Data Management in the Big Data Era

Martha O. Perez-Arriaga(✉) ⓘ and Krishna Ashok Poddar ⓘ

Veterans Affairs Cooperative Studies Program, Clinical Research Pharmacy
Coordinating Center, Albuquerque, NM 87106, USA
Marthao.pa@gmail.com, Krishna.apoddar@gmail.com

Abstract. The Department of Veterans Affairs (VA) Cooperative Studies
Program (CSP), Clinical Research Pharmacy Coordinating Center (Center) has
supported clinical trials for more than four decades. Managing information from
clinical trials and published results in the Big Data era presents new challenges
and opportunities. These include and are not limited to data attribution, aggregation, adaptability, and prompt analysis. Hence, the Center has created a
dynamic application to present a broad understanding of the clinical trials'
achievements. To collect crucial information from clinical trials, this application
includes 1) data attribution to identify provenance and to preserve relationships
between trials and resulting publications, 2) data normalization to deal with
variety of formats and concepts, 3) data aggregation to integrate information
from different trials, and 4) data analysis with a friendly interface to consult
aggregated information promptly. This work establishes a Semantic Data Model
for each clinical trial to create a summary of key information in a machine-
readable format, and to enrich each summary with semantic information. In
addition, it allows the union of these models to represent a global knowledge
source from a set of clinical trials. The organized models offer compatibility and
interoperability within and among clinical trials.

Keywords: Clinical trials · Clinical data management systems · Semantic data
model · Knowledge representation · Big data modeling · Big data applications ·
Big data analytics

1 Introduction

1.1 Clinical Trials at the Cooperative Studies Program

The Department of Veterans Affairs (VA) Cooperative Studies Program (CSP) is a
clinical research program that provides definitive answers to important clinical questions impacting the Veteran through conducting multicenter clinical trials [1]. The CSP
has conducted trials since the nineteen fifties and expanded in the early seventies with
the addition of coordinating centers. The CSP currently consists of five Biostatistical
and Data Management Coordinating Centers (BDMCCs), the Clinical Research
Pharmacy Coordinating Center (Center), five Epidemiological Centers, a Pharma-
cogenomics Laboratory, and a Network of Dedicated Enrollment Sites [2]. Research
comprises health care areas such as mental health, cancer, cardiology, and diabetes,

© Springer Nature Switzerland AG 2020
S. Nepal et al. (Eds.): BIGDATA 2020, LNCS 12402, pp. 190–205, 2020.
https://doi.org/10.1007/978-3-030-59612-5_14

among others. The Center and the BDMCCs collaborate closely to coordinate clinical trials. These trials have yielded pivotal findings, such as "Combined angiotensin inhibition for the treatment of diabetic nephropathy" [3] and "A vaccine to prevent herpes zoster and postherpetic neuralgia in older adults" [4].

To coordinate clinical trials, the Center implements a quality program called Total Integrated Performance Excellence System (TIPES) [2]. TIPES includes the Regulatory and Patient Safety Management to comply with regulations and ensure patients' safety during clinical trials; the Clinical Materials Management to manufacture, package, label, distribute, and manage clinical items, such as drugs and devices; and the Pharmaceutical Project Management to manage pharmaceutical aspects and participate in writing protocols to define a trial's design, details and approach. Finally, the Performance Excellence Management promotes innovation and continuous improvement. The Center collaborates with BDMCCs, investigators, and research coordinators in the U.S. and other countries. To manage the progress of clinical trials, the Center creates customized software systems to ensure data management activities' audit trail, research regulations, and participants' safety within a quality framework.

1.2   Challenges to Manage Clinical Trials in the Big Data Era

The main goal of conducting trials is to improve the lives of people. Hence, most trials aim to find novel and useful information to eliminate or lessen an adverse health condition. Clinical trials generate two kind of outputs: clinical results and scientific publications. A completed trial produces a variable number of publications with detailed information. Some publications cite a trial within free text as an alternative to using the references section. Others cite its source trial in the special section "trial registration" using an identifier [5]. In turn, the resulting scientific publications are cited by other researchers. Therefore, it is difficult to keep track of an accurate number of publications citing a clinical study. A current search for "Clinical Studies Cooperative Studies Program" at Google Scholar produces more than 1.6 million results. If we consider the number of connections among trials and publications, the connections increase exponentially. The large number of disconnected clinical trials and publications hinder tracking and reusing their information.

Resulting publications add value to a completed clinical trial because they present detailed findings. However, a scientific publication can lose relevant information from its main source once it is released to the Internet. Hence, there is a need to keep publications linked to their source clinical trials. It is important to establish a process to preserve attribution with minimal information, i.e. metadata to identify relevant information from clinical trials and their scientific publications. This attribution increases the value of trials data because it preserves reliable information. Attribution is also important due to proliferation of distorted information on the Internet. Therefore, linking scientific results to its original clinical trial helps identify accurate information and reliable institutions for possible collaboration.

Sharing clinical trials information is promoted, even though it has challenges and costs associated [6]. Repositories share clinical data around the world [7, 8]. Sharing clinical data remains difficult because of a) sensitive data protection, b) proliferation of data representations and management methods, c) data variety, and d) lack of a structured organization to preserve trials data with published results. Research organizations closely observe privacy constrains to protect sensitive information and preserve data integrity. Thus, confidential information cannot be shared. On the other hand, the shareable public information provides insights to determine novel, similar, and even opposite research among institutions. Hence, sharing compatible information is crucial to allow interoperability within and among institutions.

The U.S. National Library of Medicine recognizes the importance of open information from clinical studies. The Food and Drugs Administration defines normativity to publish information in clinical databases [9]. In particular, the site ClinicalTrials.gov collects information that includes results, health conditions, drugs, biologic products, behavioral intervention, surgical procedures, and medical devices. Research organizations must report required information, including clinical and published results for every study as established in [10]. To date, Clinicaltrials.gov [7] collects relevant information from more than 338,000 clinical studies in 211 countries, including the United States. Similarly, the European Union (EU) Clinical Trials Register database comprises more than 55,000 studies in the EU and other locations [8].

Before results are published in ClinicalTrials.gov, investigators submit clinical trials information to the National Library of Medicine, National Institutes of Health. A standard submission process for basic results exists [10]. The information to report includes participant flow, baseline characteristics, outcome measures with statistical analyses, and adverse events. ClinicalTrials.gov offers an open interface to search for clinical trials with standardized information. In addition, the site offers an option to download data from a search, either into a single file with all trials or an individual file per trial. Every trial can be consulted online. To review and analyze trials globally, a systematic method to aggregate this information is required.

Several institutions have worked to facilitate data aggregation of clinical trials and to improve data standardization. The National Center for Biomedical Ontology hosts BioPortal [11], a repository to comprise biomedical ontologies. The ontologies can be accessed online [12]. The Ontology of Clinical Research offers modules to represent entities from human studies [13]. On the other hand, the Biomedical Research Integrated Domain Group establishes a comprehensive model with defined concepts for clinical research [14]. These efforts are relevant for integration and include specific information from clinical trials outcomes, especially clinical results.

Variation of names for clinical concepts and formats to represent data make it difficult to integrate clinical trials information. For instance, a trial refers to a health condition as "stone in bladder diverticulum" and another as "calculus in diverticulum of bladder." Different names and acronyms of same condition produce inconsistent results for automated aggregation. Hence, normalization of data and formats would facilitate integration. A single institution may use a variety of formats to represent trials data. The Clinical Data Interchange Standards Consortium (CDISC) establishes standards for study design, data collection, data exchange, analysis, reporting, and archive [15]. ClinicalTrials.gov uses the Clinical Trial Registry XML (CTR-XML) standard from CDISC in XML (eXtended Markup Language) format [16].

Techniques for data management pose another variation. Most institutions manage information using Structured Query Language (SQL) or Not Only SQL databases [17]. These techniques have advantages and disadvantages as detailed in [18]. Preferring one data management relies on the characteristics of an institution and its data. Other factors to influence data management include technical knowledge and legacy applications. The latter requires expertise and computer resources to migrate data to new software applications. A model, compatible to any data management technique, would be useful to facilitate clinical trials data management from a global perspective.

Data management of clinical trials and their published results present challenges that include connecting a completed trial to its scientific articles; maintaining reliable and valuable data; sharing public information and protecting sensitive data; managing a large volume of resulting data; unifying a variety of data, formats, and naming conventions for clinical concepts; selecting management techniques; and integrating data from diverse sources for further analysis. Therefore, this work focuses on solving some of these issues. To do so, it uses a model to preserve relevant information from each trial and a software application to manage the resulting models.

1.3   Application to Manage Clinical Trials Information

To solve some challenges for clinical trials data management in the Big Data era, we develop an application that constructs an organized data model to summarize each clinical trial. This model tracks minimal, yet relevant identification data, i.e. metadata, to connect information from trials and results. It also contains standard data among studies and semantic information. Thus, it is called a Semantic Data Model. The models include context to identify trials topics and normalized information to avoid ambiguity. Hence, it facilitates automated data integration. The integration of models from clinical trials produces a global source of information related to trials, publications, health conditions, and even sponsors. Finally, the integrated data facilitates comparison among trials from diverse institutions. Section 2 describes the components of the Semantic Data Model to preserve relevant information of a clinical trial.

The structure of these models is compatible with the Semantic Web [19]. The Semantic Web technologies facilitate interoperability not only between people and machines, but also among organizations. Semantic web tools, such as ontologies and controlled vocabularies, normalize data for analysis and interaction with a myriad of clinical trials around the world. To alleviate naming variety, we use context and an ontology to disambiguate concepts with more than one name. An established ontology enriches concepts with semantic information and facilitates data understanding. In addition, a controlled vocabulary identifies context per trial. Section 3 describes the methods to detect context and disambiguate concepts in each model.

Finally, we present a dynamic application to manage Semantic Data Models that summarize clinical trials. This application allows users to understand, find, and explore data associations among clinical trials. Section 4 describes this application features, a use case, and design of experiments to evaluate normalization of concepts. Section 5 reports this application's usability to analyze information from a set of clinical trials. Section 6 describes conclusion and future work. Our main contribution includes an application 1) to create a Semantic Data Model that summarizes each clinical trial with metadata and facilitates traceability and reusability of information, 2) to normalize clinical concepts using semantic tools, and 3) to manage and integrate models with relevant trials' information to facilitate interoperability.

1.4   Related Work on Clinical Trials Data Management

We present a brief review of work to manage and represent clinical trials data. Krishnankutty et al. describe a summary of Clinical Data Management Systems (CDMS): commercial and open source [20]. Commercial software to manage clinical trials information include Oracle Clinical [21], ClinTrialWorks [22], and IBM Clinical Development [23]. Commercial software offers useful options to manage trials. These, however, may be expensive. On the other hand, open source software includes OpenClinica [24] and OpenCDMS [25].

To manage the development of clinical trials, the Center relies on Inventory Tracking System, Randomization and Treatment Assignment System, and Participating Pharmacy System [2]. These customized systems manage data based on each trial's protocol. Even though these systems communicate to each other, each trial remains independent of others because they do not require to exchange information. The BDMCCs manage clinical data per trial. In this work, we create a data model with semantics to summarize each trial and aggregate public information from different trials.

Data representation of clinical trials includes online repositories [7, 8] and portable files [7, 26]. Organizations rely on Structured Query Language (SQL) and Not Only SQL databases to store information [18]. Tasneem et al. create an aggregated database from ClinicalTrials.gov for analysis [27]. Their work classifies studies by clinical specialties until 2010. This database has been upgraded until 2017. Our work uses a model to manage clinical trials data and can be implemented using SQL or NO SQL. This work also offers an automated process to import data as needed.

Different standards to represent clinical data exist. ObTiMA imports and exports clinical data using the CDISC Operational Data Model in XML format [26]. The site ClinicalTrials.gov uses the CTR-XML standard to represent clinical data. Our work imports standardized information from ClinicalTrials.gov and can manage private data within an institution. In addition, it exports data models from trials in an organized JavaScript Object Notation (JSON) format. This organization describes trials information compatible with the class "Medical Trial" from Schema.org, a vocabulary to represent classes with common concepts on the Internet [28].

Ontologies and vocabularies represent important sources of knowledge. An ontology contains well defined entities, i.e. concepts. For instance, the entity *Study Type* has instances *Quantitative* and *Qualitative*. In turn, a *Quantitative* study has instances *Interventional* and *Observational*. It is common to find various ontologies in each area of study. Hence, choosing or developing a proper ontology is complicated. Fung and Bodenreider point out the proliferation of ontologies and difficulties mapping them in a single domain [29]. The Human Disease Ontology collects health conditions [30] and we use it to enrich conditions with semantics. This ontology contains cross reference information from the controlled vocabularies Unified Medical Language System (UMLS) [31] and Medical Subject Headings (MESH) [32]. Similarly, it contains the 10th revision of the International Statistical Classification of Diseases and Related Health Problems (ICD10) to provide codes for diagnosis.

This application builds a model to represent a clinical trial and its publications metadata. A model is a combination of information from ClinicalTrials.gov and articles metadata as used in the semantic data models from scientific publications [33]. The structured organization facilitates compatibility and interoperability among trials. ClinicalTrials.gov contains information from interventional, i.e. clinical trials, and observational studies. Even though we focus on the former, the model can handle data from both kind of studies. The following section describes the components of a Semantic Data Model for clinical trials.

2 Semantic Data Model of Clinical Trials

2.1 Semantic Data Model of a Clinical Trial

We propose a Semantic Data Model to summarize the most relevant information from a clinical trial and facilitate data aggregation from clinical trials. To do so, metadata and context represent important information and identify trials. Metadata simplifies information management, connects scientific publications to original trial, and increases trials' traceability and reusability. Context recognizes the interest and area(s) of study in a trial to simplify analysis. This model unifies the variety of data, formats, and naming conventions for clinical concepts. Similarly, the model can be implemented using different data management techniques. Hence, the model facilitates aggregation, attribution, and analysis of trials.

2.2 Elements of a Semantic Data Model

The Semantic Data Model defines the main components of a clinical trial: metadata and context; details; design; data related to each trial's approach, including health conditions, devices, and drugs; published results and related publications, and institutions that coordinate trials, such as sponsor and collaborators. Figure 1 shows the main components of this Semantic Data Model. Every element of this model contains metadata. To describe the context of trials, this model uses keywords. Finally, this model uses a machine-readable format that can interact with the Semantic Web.

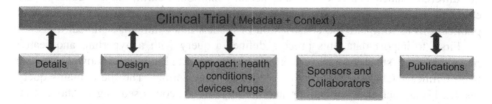

Fig. 1. The main components of a semantic data model summarize a clinical trial.

To ensure reproducibility and collect metadata, we use the ClinicalTrials.gov repository [7] and CTR-XML standard [16]. Details comprise trial and participants information. A trial includes numeric identifier, ClinicalTrials.gov identifier, title, short title, description, phase, status, type, start and completion dates, and keywords to describe its context. Details also include number of participants per trial, enrollment criteria, and general characteristics, i.e. age and sex. Design consists of allocation, intervention model, and masking. Design concepts describe how participants are grouped and treated in a trial, e.g. randomized allocation.

Every trial indicates an approach to fulfill its goal. A trial targets a health condition using drug or other therapies. For instance, a trial focuses on improving Diabetes Management using an electronic device to measure physical activity. Then, this trial contains information related to health conditions and devices. For now, we focus on describing health conditions. To do so, we include information from the Human Disease Ontology [30], such as identifier, name, definition, synonym, class, and a Uniform Resource Identifier (URI) to indicate a medical guideline. It also includes equivalent identifiers for same health condition at controlled vocabularies, such as ICD10, MESH and UMLS. Keywords from the MESH vocabulary define each trial's context.

This model emphasizes published results to relate and trace them with its original trial. Hence, publications from ClinicalTrials.gov are collected. They include publications from completed trials, publications reported by trials data providers, and publications identified by Medline (a digital database of health information). Each publication's reference appears as text. To identify a reference, we classify this text as metadata, e.g. identifier, title, trial identifier, authors, and a Digital Object Identifier (DOI). Finally, each model contains information of institutions responsible to coordinate research, such as sponsor and collaborators.

2.3   Construction of a Semantic Data Model

Automated steps to construct a Semantic Data Model of a clinical trial include: 1) data import, 2) data extraction and context detection, 3) data normalization, 4) semantic enrichment, and 5) metadata extraction of published results. To automate this process and make it reproducible, we use the repository ClinicalTrials.gov. First, a query selects the trials of interest from this reliable repository as part of the data import process. Second, we extract data to populate our models and detect trials' context. Third, we use a Natural Language Processing method to normalize clinical concepts, i.e. different names that refer to same health condition. Fourth, clinical concepts are enriched with definitions of a maintained ontology. Fifth, we extract metadata of publications to complete the construction of models that summarize clinical trials.

First, to import data, this process defines a query with a hyperlink and search parameters, e.g. studies by a sponsor. Then, this hyperlink is used to download a file from ClinicalTrials.gov in a comma separated value format. The file contains query results. Using each trial's identifier, another hyperlink is composed to get data per trial in XML format. This data contains the elements of a model: details, design, approach, sponsors, and collaborators. Similarly, an additional hyperlink per trial is composed to get publications information. Hence, we use a set of hyperlinks to download documents with trials of interest.

Second, to extract data, each XML document undergoes a processing stage to identify specific data for a model. We redeem a unique identifier per keyword. Keywords, defined with tag <mesh_term>, represent a context in each trial. To avoid storing a keyword multiple times, this process assigns an identifier when a new keyword is encountered. Similarly, we use <allocation>, <intervention_model>, and <masking> for design; <lead_sponsor> and <collaborator> for agencies; and other tags to extract approach and details of each trial. These tags extract health conditions, primary study, title, brief title, age, enrollment, exclusion criteria, healthy volunteers, inclusion criteria, gender, phase, status, description, and start and completion dates..

Third, to normalize names, we use health conditions from each clinical trial's approach. Because there exist different naming conventions for same concept, we automate data normalization to identify unique names. Unique keywords are used as a trial context. We download the Human Disease Ontology [34] to detect a unique identifier for each health condition in a trial. Fourth, to enrich semantically a health condition, we use the identifier from previous stage to detect its definition, equivalent identifiers in other knowledge sources, and a Uniform Resource Identifier to represent its medical guideline (see Sect. 3).

Finally, to preserve the relationship between a trial and its published results, this process links each resulting publication with its original clinical trial in a model. From the ClinicalTrials.gov files, we use the tag <reference> to extract metadata of each publication. The reference contains free text and lacks tags to identify its elements, e.g. title. Then, this text undergoes a processing to separate metadata from each reference. Regular expressions separate reference elements, i.e. title and author name(s). This process assigns identifiers to each publication and author for further identification and organization.

After this processing, we collect a Semantic Data Model for each clinical trial. Its representation contains semantic information from an ontology and controlled vocabularies. This model can be adopted in a SQL database or Not only SQL database. A composed Semantic Data Model can be exported in a JSON format. Given that metadata identifies details of a clinical trial and its results, each model represents a summary per study. This model facilitates storing provenance of information and keeps associations of a trial with its design, participants characteristics, published results and authors, health conditions, and coordinating agencies.

3   Semantic Enrichment for Concepts in Clinical Studies

3.1   Context in Clinical Trials

The description of a trial explains its characteristics and objectives. In general, a context is understood from this description. However, defining a context in each clinical trial facilitates automated identification of trials with similar and opposite objectives. A context is key to aggregate information and it can be represented with keywords. Most trials at ClinicalTrials.gov contain keywords using the MESH vocabulary. This approach takes advantage of keywords defined in a study and uses them as a context. These keywords help normalize health condition names in each clinical trial.

3.2   Health Conditions

A Natural Language Processing method is used to match an identifier for each trial's health condition. This process finds a unique identifier from the Human Disease Ontology to normalize a concept. Before this process and for a single time, we create an inverted file index to sort concepts from this ontology, as explained in [35]. An inverted file index contains each concept with a list of health conditions identifiers. Identifiers are recorded when a concept exists in a health condition name, synonym, or definition within this ontology. The inverted index stores the Human Disease Ontology with 12,073 health conditions and more than 18,100 concepts in a JSON file. Later, this index facilitates finding a concept mentioned in a trial's condition.

We perform Latent Semantic Analysis [36], a Natural Language Processing to find a unique identifier per health condition in a trial. A search vector V_S represents a health condition. To create V_S, we use text from a clinical trial's title, description, context, and name of a health condition. This textual information undergoes a normalization step to eliminate stop words and generate unique tokens with their frequency. From this normalization, we obtain V_S. Then, the previously created inverted index serves to rapidly search for concepts from a health condition name and a trial's context. Each condition identifier found in the inverted index acts as a candidate to find a normalized health condition name.

Using the candidate identifiers, we create a matrix M_D of vectors. M_D reduces dimensionality to search for the right identifier. To select parameters, we previously examined bag of words and n-grams ($n = 2,3$). Then, we create vectors with two grams and 150 features. Each vector in M_D undergoes the same normalization step as V_S. This process calculates similarity between set of vectors M_D versus vector V_S. Thus, this process selects the most similar vector from M_D to V_S. This comparison gets the closest semantic description for a trial's health condition and its identifier from the Human Disease Ontology. In addition, this ontology enriches the model with semantic information per health condition in a clinical trial.

4   Application to Manage Semantic Data Models

4.1   Application Features and Use Case

This application has twofold goals: to present an overview of clinical trials with published results and to prepare data to effortlessly interact with other trials in the Big Data era. The Internet provides a wealth of data and processing is necessary to find specific information. This application identifies, integrates, and analyzes trials information. In addition, the application supports researchers to get specific answers.

Before planning a clinical trial, researchers analyze and collect related publications. In addition, they compare data from similar completed and ongoing trials. Researchers require answers from studies supported by all or specific sponsors. As use case examples, a researcher requires to answer the following questions: 1) what is the most used design for clinical trials? 2) what is the number of participants in completed studies? 3) what studies have investigated conditions related to liver? and 4) what publications are related to such studies? The collection of information depends on accessible resources and manual tasks that can be time consuming. Time is subjective to each research question.

This application can collect all public studies or a filtered set of studies. For instance, it collects studies supported by a specific sponsor. For this demonstration, we select one thousand studies with sponsor VA Office of Research and Development (ORD) from ClinicalTrials.gov. To show the usability of this application, Semantic Data Models are created with automated steps previously described (see Sect. 2.3). To create, manage, and integrate semantically enriched data models, this application contains three main modules: 1) *data model*, 2) *overview*, and 3) *analysis*.

The first module *data model* comprises the options a) build, b) normalize, and c) export. The *build* option extracts data from select trials at ClinicalTrials.gov and populates information into data models. Once the models are built, we opt to use a SQL management approach. Then, the *normalize* option uses the models to find unique identifiers for each health condition in a trial. In addition, this option enriches the models with semantic information from the Human Disease Ontology. The *export* option in this module generates a JSON file with the created models. We use this export file to transfer clinical trials information into a Not Only SQL approach.

The second module *overview* displays summaries from the set of models created in previous module. This overview shows data for a) clinical studies and b) articles, including published results and related publications. From the set of collected *clinical studies*, users can view a list of all trials with its identifier, title, sponsor, and health conditions. If users require to view more information per study, the application directs them to ClinicalTrials.gov. The *articles* option shows titles of each published result with its trial identifier and authors.

Finally, the third module *analysis* presents statistics and aggregated data using the generated models. It includes a) context, b) design, c) phase, d) status, e) health conditions, and f) population characteristics. The *context* option displays studies that share context, i.e. keywords. Keywords include conditions, drug ingredients, symptoms, human organs, and other concepts related to each trial. The *design* option shows a summary of design types, number of studies, and enrollment per type. The *phase* and *status* options show a summary of the number of studies and enrollment per phase and status, respectively. Similarly, the *population characteristics* option shows the number of studies and enrollment per sex group. The *health conditions* option summarizes the number of studies per condition. Studies focus on one or various health conditions.

To evaluate some features on this application, we design two experiments to measure the normalization of naming health conditions. A random set of 60 studies and true values from the Human Disease Ontology are evaluated. The first experiment creates search vectors using title, description, context, and health conditions names with acronyms. The second experiment omits acronyms. For instance, PTSD is used in the first experiment, and replaced by Post-Traumatic Stress Disorder in the second one. Then, we calculate F1 score for both experiments.

5 Results and Discussion

5.1 Use Case Results

For demonstration purposes, we used the application to build one thousand semantic data models sponsored by the VA ORD. From these models, 101 denoted observational and 899 interventional studies, i.e. 89.9% clinical trials. The Human Disease Ontology contained 12,073 conditions that were used to normalize a variety of health condition names. From those, 7,184 conditions referred to a URI as a medical guideline, e.g., http://en.wikipedia.org/wiki/Gulf_War_syndrome for the Persian Gulf Syndrome. In addition, the first experiment using acronyms to normalize health conditions names yielded an F1 score of 0.46. On the other hand, the second experiment with no acronyms yielded an F1 score of 0.74.

In the *overview* module, we obtained one thousand studies with clinical trial identifier, title, study type, health condition(s), sponsor, and view button to direct users to ClinicalTrial.gov. There were 347 studies with 1,449 published results. From those, 254 were completed studies with 753 publications. The 93 remaining non-completed trials contained 696 publications, which were reported by data providers or Medline. The application also identified 442 authors from the 1,449 publications.

In the *analysis* module, the application collected studies per keyword. Keywords represented context per trial and 1,483 unique keywords were detected. We found 67 different kinds of design. The design most used by 199 trials was the combination of allocation: randomized, intervention model: parallel assignment, and masking: single (outcomes assessor). The second design with 180 studies was randomized, parallel assignment, and none (open label). The third design with 77 trials uses intervention model: single group assignment and masking: none (open label). For use case, this result answered researcher's first question, most common design in clinical trials set.

The *analysis* module found 26 trials for phase 1, 84 for phase 2, 34 for phase 3, and 43 for phase 4. In addition, we found 644 trials with no applicable value, and 101 observational studies with no value. For status, we found 93 active (not recruiting) studies, 495 completed, 29 enrolling by invitation, 73 not yet recruiting, 272 recruiting, two suspended, 28 terminated, one unknown status, and seven withdrawn studies. There were 495 completed studies with 1,452,427 participants. For use case, this answered researcher's second question, number of participants in completed studies by this sponsor. Table 1 shows number of studies and enrollment per status.

In addition, the *analysis* module found 727 health conditions in this set of studies. There were 138 trials that targeted post-traumatic stress disorder, 61 for depression, 60 related to diabetes, 54 for stroke, 45 related to pain, 42 for schizophrenia, 42 for traumatic brain injury, 36 for spinal cord injury, 32 related to cancer, 29 for Parkinson's disease, 25 for obesity, 21 for dementia, 19 for hypertension, and 18 related to aging, among others. Finally, we found the total number of studies and participants enrolled per sex group (see Table 2). From the more than two million participants, 201,294 were from observational studies.

Table 1. Number of studies and enrollment per status.

Status	Number of studies	Enrollment
Active, not recruiting	93	171,312
Completed	495	1,452,427
Enroll by invitation	29	12,815
Not yet recruiting	73	24,650
Recruiting	272	367,076
Suspended	2	74
Terminated	28	1,612
Unknown status	1	20
Withdrawn	7	20

Table 2. Number of studies and enrollment per sex group.

Sex	Number of studies	Enrollment
All	946	2,002,156
Female	17	18,755
Male	37	9,075
Total	1,000	2,029,986

To answer the researcher's third and fourth questions from use case examples, the module *analysis*, option context, showed studies NCT00252499, NCT00108589, NCT01289639, and NCT01002547 related to fatty liver; NCT04178096, NCT03796598, and NCT00108355 to liver cirrhosis, and NCT04166240 to liver neoplasms. These studies targeted different health conditions, e.g. hepatitis, non-alcoholic steatohepatitis, fatty liver, and insulin resistance. Finally, the module *overview* showed title, authors, and DOI of 15 publications related to these liver studies.

5.2 Discussion

Organizations preserve old clinical trials' data, not found in public repositories. On the other hand, ClinicalTrials.gov provides a wealth of open information to create data models. Accordingly, organizations can use our presented methods to manage, integrate, and compare clinical trials data from diverse sources.

ClinicalTrials.gov requires reporting clean data. Using ontologies to define data elements in a clinical trial, from its planning stage, improves reporting data. Reporting health conditions from an ontology minimizes the need of data normalization. Similarly, mandatory definition of clinical trials' context improves data aggregation. Naming conventions to report data, e.g. sponsor and title, should be standardized for better data management in global repositories and organizations. For instance, to normalize a clinical research title, the Coordinate Studies Program uses acronym CSP, identifier, and descriptive title: CSP #2008 – Pentoxifylline in Diabetic Kidney Disease.

We developed automated methods to create models from clinical trials data. To enhance models with semantics, we performed data extraction and normalization. Normalization is key for data aggregation. Evaluating normalization, our approach found correct names of health conditions using context and description of a clinical study. This approach misclassified some conditions. It can be attributed to lack of a trial's context or description with unrelated details to condition. Acronyms, common in the Health Sciences, hindered finding the right condition. Replacing acronyms by words improved the approach and F1 score from 0.46 to 0.74. These results, for matching correct health conditions' identifiers and semantics, were fair. A probabilistic approach will improve further the data normalization process.

There were concepts not found in the ontology used, e.g. "chronic diabetic foot ulcers." Our approach matched this trial to disease "chronic ulcer of skin." Thus, an expert should review and accept results if appropriate. The ontology contained some obsolete condition name(s) repeated within new names. Some clinical trials focus on concepts that are not necessarily a health condition, e.g. quality of life, elderly adults, and job satisfaction. Then, other concepts require normalization. Besides automated methods, collaboration among institutions producing ontologies and research coordinators benefits ontologies' development and completeness.

Attribution of publications to its original trial was performed within the models. We linked 1,449 publications with their original trial and traced other relationships, e.g. authors. However, not all publications result from trials. Some of them include design articles and related work. Reporting each publication under its correct category, i.e. design, result, and related, facilitates data attribution. Otherwise, we require a classification process to identify each publication kind. In addition, citing clinical trials formally, in the references section of scientific publications, facilitates tracking resulting publications from trials.

This work's analysis module identified statistics. For instance, it found 67 different design types to implement trials. This work also identified enrolled participants and studies per status (see Table 1). It identified studies per phase and studies sharing similar context and health conditions. More than two million people have participated in these studies (see Table 2). Approximately, 90% participants were from interventional and the rest from observational studies. In this set, there were 17 studies for only female group and 37 for only male. Female enrollment was higher than male participants. The 946 studies including all gender had the highest enrollment.

The models generated were analyzed in different databases, i.e. SQL and No SQL. Because the models include relevant organized information, the application supports further data analysis, e.g. enrollment by sponsor, studies by age range, and publications by author. This application created models to manage data and prepare trials information for traceability, compatibility, and interoperability. This proof of concept showed that semantic data models facilitate these tasks and provide specific answers.

6 Conclusion

6.1 Lessons Learned and Future Work

We studied how clinical trials data management fits in the Big Data era. Clinical trials data include issues common to other areas: dealing with data variety, managing their volume of results, protecting sensitive data, and maintaining the value and veracity of information. To solve some of these issues, we created a Semantic Data Model to organize and summarize information from a clinical trial. Semantic information was deemed important for data management. Hence, semantic tools, such as vocabularies and ontologies, were used to enrich concepts. Natural language processing normalized names of health conditions. A model supported information from interventional or observational studies. Models linked published results to trials. Each model contained unique characteristics organized in a machine-readable format. This organization facilitated data management using SQL and NoSQL approaches. Finally, the aggregation of models analyzed information from a global perspective.

One of the applications of aggregating clinical trials is to reuse information of trials already performed. Different data elements helped aggregating information. Context was crucial to integrate information and find relationships between studies. Automated methods aggregated a variety of data. This aggregation made it possible to find different research groups with similar targets in clinical trials, e.g. health conditions. The collected information in models facilitated standardization and identified statistics from studies. The models built from this application offer reusability of information and possibilities of collaboration for ideas exchange.

Public repositories are valuable for research organizations. Reporting normalized data increases research value. Clinical trials investigate and generate data from health conditions, drugs, and medical devices. A growth of trials for hybrid studies with multiple targets, e.g. genetic and sensors data, is expected. These trials require specialized ontologies to organize information. Active collaboration between ontologies developers and clinical researchers benefits naming conventions with semantic data.

Visualization is paramount in the Big Data era because it supports planning, analysis, and development of clinical trials. Then, a visualization module for this application will facilitate to explore data associations. In addition, we aim to detect close relationships among studies from different organizations using a large set of studies. In our connected world, data definition, attribution, normalization, and integration remain important for better clinical trials data management.

Acknowledgments. This research was supported in part by the Department of Veterans Affairs, Veterans Health Administration, Office of Research and Development, Cooperative Studies Program using resources and facilities at the VA Cooperative Studies Program Clinical Research Pharmacy Coordinating Center. The authors thank Zachary Taylor, Kathy Boardman, Heather Campbell, and anonymous reviewers for valuable comments to improve this manuscript. Similarly, we acknowledge Todd A. Conner for his encouragement and relevant comments to improve this work.

Disclaimer. Contents are expressed by the authors and do not represent the views of the Department of Veterans Affairs or the United States Government.

References

1. Henderson, W., Lavori, P., Peduzi, P., Collins, J., Sather, M., Feussner, J.: Methods and Applications of Statistics in Clinical Trials: Planning, Analysis, and Inferential Methods, vol. 2. Chapter 55, U.S. Department of Veterans Affairs Cooperative Studies Program. Wiley (2014)
2. Sather, M., et al.: Total integrated performance excellence system (TIPES): a true north direction for a clinical trial support center. Contemp. Clin. Trials Commun. **9**, 81–92 (2018)
3. Fried, L.F., et al.: VA NEPHRON-D investigators: combined angiotensin inhibition for the treatment of diabetic nephropathy. New England J. Med. **369**(20), 1892–1903 (2013)
4. Oxman, M.N., et al.: A vaccine to prevent herpes zoster and postherpetic neuralgia in older adults. New England J. Med. **352**(22), 2271–2284 (2005)
5. Yakovchenko, V.: Automated text messaging with patients in department of veterans affairs specialty clinics: cluster randomized trial. J. Med. Internet Res. **21**(8), e14750 (2019)
6. Homepage Committee on Strategies for Responsible Sharing of Clinical Trial Data; Board on Health Sciences Policy; Institute of Medicine. Sharing Clinical Trial Data: Maximizing Benefits, Minimizing Risk. Washington (DC): National Academies Press (US); 2015 Apr 20. 6, The Future of Data Sharing in a Changing Landscape. https://www.ncbi.nlm.nih.gov/books/NBK285998/. Accessed 29 Jan 2020
7. Clinicaltrials.gov Homepage. https://clinicaltrials.gov. Accessed 08 May 2020
8. EU Register Trial Homepage. https://www.europeandatajournalism.eu/eng/News/Useful-data/EU-Clinical-Trials-Register. Accessed 08 May 2020
9. Hyman, P., McNamara, P.C.: Food and Drug Administration Amendments Act of 2007 (2007)
10. Tse, T., Williams, R.J., Zarin, D.A.: Reporting "basic results" in ClinicalTrials. gov. Chest **136**(1), 295–303 (2009)
11. Whetzel, P. L., et al.: BioPortal: enhanced functionality via new Web services from the National Center for Biomedical Ontology to access and use ontologies in software applications. Nucleic Acids Res. **39**(suppl_2), W541–W545 (2011)
12. Salvadores, M., Alexander, P.R., Musen, M.A., Noy, N.F.: BioPortal as a dataset of linked biomedical ontologies and terminologies in RDF. Semant. Web **4**(3), 277–284 (2013)
13. Sim, I.: The ontology of clinical research (OCRe): an informatics foundation for the science of clinical research. J. Biomed. Inform. **52**, 78–91 (2014)
14. BRIDG Model. https://cbiit.github.io/bridg-model/HTML/BRIDG5.3.1/. Accessed 12 Dec 2019
15. Souza, T., Kush, R., Evans, J.P.: Global clinical data interchange standards are here! Drug Discovery Today **12**(3–4), 174–181 (2007)
16. CDISC Standards Data Exchange. https://www.cdisc.org/standards/data-exchange. Accessed 10 Dec 2019
17. Cattell, R.: Scalable SQL and NoSQL data stores. ACM Sigmod Rec. **39**(4), 12–27 (2011)
18. Wang, X., Williams, C., Liu, Z., Croghan, J.: Big data management challenges in health research—a literature review. Brief. Bioinform. **20**(1), 156–167 (2019)
19. Matthews, B.: Semantic web technologies. E-learning **6**(6), 1–19 (2005)
20. Krishnankutty, B., Bellary, S., Kumar, N.B., Moodahadu, L.S.: Data management in clinical research: an overview. Indian J. Pharmacol. **44**(2), 168–172 (2012)
21. Oracle Clinical. http://www.oracle.com/us/industries/life-sciences/045788.pdf. Accessed 09 Dec 2019
22. ClinTrialWorks. https://www.clintrialworks.com/. Accessed 13 Dec 2019

23. IBM Clinical Development. https://www.capterra.com/p/139887/IBM-Clinical-Development/. Accessed 09 Dec 2019
24. OpenClinica. https://www.openclinica.com/open-source-clinical-trial-software/. Accessed 09 Dec 2019
25. OpenCDMS Homepage. https://www.medfloss.org/node/408. Accessed 09 Dec 2019
26. Stenzhorn, H., et al.: The ObTiMA system-ontology-based managing of clinical trials. Stud. Health Technol. Inform. **160**(Pt 2), 1090–1094 (2010)
27. Tasneem, A., et al.: The database for aggregate analysis of ClinicalTrials.gov (AACT) and subsequent regrouping by clinical speciality. PLOS One **7**(3), e33677 (2012)
28. Class "MedicalTrial" from Schema.org. https://schema.org/MedicalTrial. Accessed 12 Dec 2019
29. Fung, K.W., Bodenreider, O.: Knowledge representation and ontologies. In: Richesson, R., Andrews, J. (eds.) Clinical Research Informatics. Health Informatics, pp. 313–339. Springer, Cham (2019). https://doi.org/10.1007/978-1-84882-448-5_14
30. Schriml, L.M., et al.: Human Disease Ontology 2018 update: classification, content and workflow expansion. Nucleic Acids Res. **47**(D1), D955–D962 (2019)
31. Bodenreider, O.: The unified medical language system (UMLS): integrating biomedical terminology. Nucleic Acids Res. **32**(suppl_1), D267–D270 (2004)
32. Lipscomb, C.E.: Medical subject headings (MeSH). Bull. Med. Libr. Assoc. **88**(3), 265–266 (2000)
33. Perez-Arriaga, M.O., Estrada, T., Abad-Mota, S.: Construction of Semantic Data Models. In: Filipe, J., Bernardino, J., Quix, C. (eds.) Data Management Technologies and Applications. DATA 2017. Communications in Computer and Information Science, vol. 814, pp. 46–66. Springer, Cham (2018). https://doi.org/10.1007/978-3-319-94809-6_3
34. Human Disease Ontology, Bioportal. https://bioportal.bioontology.org/ontologies/DOID. Accessed 16 Dec 2019
35. Amato, G., Savino, P.: Approximate similarity search in metric spaces using inverted files. In: Proceedings of the 3rd International Conference on Scalable Information Systems, pp. 1–10. ICST: Institute for Computer Sciences, Social-Informatics and Telecommunications Engineering (2008)
36. Landauer, T.K., Foltz, P.W., Laham, D.: An introduction to latent semantic analysis. Discourse Processes **25**(2–3), 259–284 (1998)

Cross-Cancer Genome Analysis on Cancer Classification Using Both Unsupervised and Supervised Approaches

Jonathan Zhou[1]([✉])(iD), Baldwin Chen[2], and Nianjun Zhou[3]

[1] Horace Greeley High School, Chappaqua, NY 10514, USA
jozhou@students.ccsd.ws
[2] Ardsley High School, Ardsley, NY 10502, USA
baldwinchen@gmail.com
[3] IBM, 1101 Kitchawan Road, Yorktown Heights, NY 10598, USA
jzhou@us.ibm.com

Abstract. Many problems exist within the current cancer diagnosis pipeline, one of which is alarmingly high over-diagnosis rates in breast, prostate, and lung cancer. Through quantifying gene expression levels, next-generation sequencing techniques such as RNA-Seq offer an opportunity for researchers and clinicians to gain a more complete view of a cell's transcriptome. With the adoption of this new data source, cross-cancer methods for cancer diagnosis have become more viable. We utilize mutual information in conjunction with a Gaussian mixture model and t-SNE to evaluate the separability of cancer and non-cancer tissue samples from RNA-Seq expression data. The Gaussian mixture and t-SNE combination produced clear clustering without supervision, suggesting the ability to separate tissue samples algorithmically. Afterwards, we use a collection of deep neural networks to classify tissue origin and status from tissue sample gene expressions. We use genes selected based on the prior mutual information technique. First, we select the top 500 genes from candidate genes without considerations for overlap in the predictability of those genes. We then applied Recursive Feature Elimination (RFE) to select 200 genes, thus accounting for covariation. We find that the performance using the top 500 genes is only slightly better than the 200 genes selected using RFE, and the two approaches achieved similar performance overall, indicating that only a small subset of genes is required for the identification of status and origin. This work indicates that RNA sequencing data is a useful tool for cross-cancer studies. Next steps include the implementation of a greater amount of non-cancer data from other datasets to decrease bias in model training.

Keywords: Cross-Cancer Genome · The Cancer Genome Atlas (TCGA) · Mutual information · Recursive Feature Elimination (RFE) · Least Absolute Shrinkage and Selection Operator (LASSO) · Gaussian mixture model · Clustering · Dimension reduction · Neural network · MLP (Multilayer Perceptron)

J. Zhou and B. Chen—Equal Contribution.

© Springer Nature Switzerland AG 2020
S. Nepal et al. (Eds.): BIGDATA 2020, LNCS 12402, pp. 206–219, 2020.
https://doi.org/10.1007/978-3-030-59612-5_15

1   Introduction

Cancer remains one of the leading causes of disease-related deaths worldwide. In 2018, the National Institutes of Health's National Cancer Institute (NCI) estimated that 609,640 patients died of the disease in the United States alone [1]. However, cancer-related deaths are much higher in countries where comprehensive cancer treatment remains a luxury.

Currently, cancer is categorized based on the tissue of origin and diagnosis is focused on the identification of stage and health of specific samples. There are several key challenges in cancer diagnosis, one of which is overdiagnosis: when an unknown tumor is classified higher grade than is present, thus encouraging severe treatments for a less severe condition. Using prostate cancer as example, researchers hypothesize that overdiagnosis rates range between 23 and 42% [2]. Prostate cancer suffers from this problem especially as a 2013 study found that biopsy Gleason scores matched prostatectomy specimen Gleason scores in only in 63% of patients and found the current 12-core biopsy method for identifying tumors to be severely underperforming [3].

This problem is not localized to prostate cancer. A study on Lung Cancer by Bhatt et al. [4] hypothesized that the very high cost involved in the complete diagnosis process, utilizing CT scans, sputum cytology, and bronchoscopy prevents accurate diagnosis of the disease in many cases. They also found that just under 80% of lung cancers are diagnosed at an advanced stage, once the disease enters a life-threatening stage.

Finally, a 2016 study found that only 30 of the 162 additional small tumors per 100,000 women that were diagnosed were expected to progress to become large, indicating that the remaining 132 cases of cancer per 100,000 women were over-diagnosed [5]. This indicates that the problem of overdiagnosis is present in a wide variety of cancers.

The development of Genomics and Data Science provides us a new angle in cancer diagnosis. Recent advanced screening devices have allowed for high throughput screening of cancer tissue with new levels of specificity. With increased levels of observable gene expression, researchers and clinicians are now able to observe greater differences in gene expression across various cancer samples, enabling us to evaluate the validity of producing computer models to differentiate across multiple cancers.

Previous sequencing research focuses much on the production and analysis of microarray data. RNA expression micro-arrays (REMs) use simultaneous hybridization with test and reference (housekeeping) genes to calculate an expression ratio based on normalization with the endogenous reference gene [5], enabling researchers to evaluate the gene expression values across various samples with a standard system of analysis.

Despite their usefulness, microarrays are fraught with several limitations. The most severe drawback is that expression values are relative to other signals detected on the array itself, and so the signal may vary from array to array.

On the other hand, next-generation sequencing (NGS) techniques based on the analysis of cDNA sequence quantities allows for quantifiable data points, offering superior accuracy in genes and complete view of genes with very high or very low expression levels [6,7]. Providing a more complete view of our gene expression enables us to develop models trained on more comprehensive data that offers a wider range with respect to gene expression, enabling a model that can be trained to differentiate tissue of origin and health via the gene expression values.

A recent study undertaken by the National Institute of Health (NIH), collected RNA sequencing data for thirty-three different types of cancer tissues. The data is compiled in a public database known as The Cancer Genome Atlas (TCGA). TCGA offer vast quantities of sequencing data with thousands of samples available for usage and analysis [8]. TCGA collects both cancer tissue and healthy (non-cancer) tissue from the same patients, thus allowing researchers to compare samples from the same body of origin. However, there is an absence of cancer free patients in TCGA.

Utilizing the vast volume and high quality of the TCGA RNA-sequencing data, we hypothesized that we could locate statistically significant genes that could influence specific cancers and identify common genes that are related to specific sets of cancers. Such a model would allow for both researchers and clinicians to predict the status of cancer tissue in question with greater accuracy than before.

The goal of our research method is to identify the type and status (cancer/non-cancer) of a given tissue sample from a limited set of gene expression data. To achieve this, we utilized Mutual Information (MI) and Recursive Feature Elimination (REF) to identify a reduced set of genes that have significance in terms of separation across both these criteria. We then employ a clustering and visualization focused approach that allows us to visualize and confirm this separation visually. This provides a directional guideline on how we should build a supervised predictive model for type and status prediction with a limited number of genes. Finally, we constructed multilayer perceptron predictive models for multiple classification across type and status.

The purpose of taking an unsupervised approach initially is to determine whether there is a natural separability of tissue origin and type with respect to gene expression from a biological perspective. The unsupervised approach provides insight into separability generally, which is indicated by both the algorithmic ability to separate the data into natural clusters and from visualization. This separability provides insight into interpretability and provides a reasonable potential clinical explanation. On the other hand, the supervised approach is taken as a method of assessing potential diagnostic accuracy and viability based off of gene expression data.

2   Related Work

While pan-cancer studies are not a new field, the use of machine learning for diagnosis across several cancers is relatively new. A recent paper published surrounding this issue focuses on the use of microarrays in the diagnosis of samples in a pan-cancer analysis. In 2018, Gao et al. [9] utilized a robust hybrid classifier, comprising of prediction analysis for microarrays and random forests to perform a pan-cancer analysis utilizing microarray data. Their model achieved relatively good accuracy and provides the groundwork for pan-cancer analysis using machine learning and indicates the viability machine learning model trained on RNA sequencing data in providing a clear and interpretable view in cancer diagnosis.

Golcuk G. et al. [10] created a model utilizing neural ladder networks and TCGA breast cancer data. Their use of neural ladder networks demonstrates the effectiveness of deep learning models in the cancer diagnosis process. Our previous research [11] studied the feasibility of utilizing machine learning models to classify prostate cancer with much fewer samples. In 2019, we utilized a logistic regression model for the prediction of prostate cancer with a SMOTE sample boosting method. Using TCGA prostate cancer genetic data, we achieved accuracies in excess of 95% for the prediction of the presence of prostate cancer. While this model focused solely on prostate cancer, the effectiveness of a machine learning model that utilized gene expression data among other features, suggests the ability of machine learning models to analyze gene expression values and classify tissue into different classes.

3 Data Preprocessing

To initiate our Pan-Cancer study, we downloaded RNA expression data for Breast (BRCA), Kidney (KIRC), Prostate (PRAD), Thyroid (THCA), Lung Squamous Cell (LUSC), and Lung Adenocarcinoma (LUAD) cancers using the TCGA Assembler 2 tool [12] in the R Statistical Computing Environment. Although there are 33 different types of cancer type in TCGA repository, only the prior six contain a reasonable number of samples from both healthy and cancer tissue. The reader can find a more descriptive figure from the Fig. 1 of Golcuk et al.'s work [10]. Each cancer type has four categories of genetic information: Gene Expression, Exon Junction, Isoform, and Exon. This study only concerns the Gene Expression portion. Gene expression type data is the most common data available and is the most cost effective to acquire.

Our dataset contains 20,531 separate features, all of which are individual genes sequenced by the TCGA project. Each feature for a given gene is measured in Reads Per Kilobase of transcript per Million mapped reads, the normalized frequencies of the gene in the RNA sequence expression data [12]. Because of this, the processed data is far more comparable from sample to sample, allowing models trained on it to more accurately analyze and predict the status of unknown tissues.

We downloaded the data for each cancer type as a separate file. For each cancer type, we mapped the healthy and cancer samples to 0 and 1 respectively. Following this, we combined the individual cancer datasets. For the combined dataset, we labeled all healthy tissue samples 0 and used a number from 1 to 6 for each of the cancer types. In total we aggregated 406 healthy and 3666 cancer samples. Table 1 provides a summary of the number of samples for different cancer types.

Table 1. Data distribution of samples for selected cancer types

Cancer type	Samples	Cancer type	Samples
Breast	1102	Lung SQ	513
Kidney	534	Prostate	502
Lung AD	517	Thyroid	498
No Cancer	406		

4   Feature Engineering and Identifying Key Genes

To determine the dependence between gene expression levels for a given cancer and the ultimate outcome of a sample (healthy/cancer), we used mutual information, a measure of the mutual dependence between the two variables, as the initial metric, calculating the mutual information between each gene and the outcome for each individual cancer type first. Mutual information between two variables is defined as follows,

$$I(X;Y) \equiv H(X) - H(X|Y)$$

expanded,

$$I(X;Y) = \sum_{y \in Y} \sum_{x \in X} p_{(X,Y)}(x,y) \log(\frac{p_{(X,Y)}(x,y)}{p_X(x)p_Y(y)})$$

where $p_{(X,Y)}(x,y)$ denotes the joint probability mass function of variables X and Y, and where p_X and p_Y denote the marginal probability mass functions of X and Y respectively. Mutual information gives a good indication of the value of a selected gene in predicting the associated cancers. We calculated a mutual information for every gene in the set with relation to all six cancer tissue types. We then selected the top 2,000 genes from the sum of the individual mutual information values across the six tissue classes. The top 11 genes with respect to this sum are displayed in Table 2.

5   Analysis of Gene Datasets with Unsupervised Learning

As a first step, we took an unsupervised approach in identifying whether there was any relation between gene features on tissue origin and type (cancer or non-cancer).

Because the dataset is extremely high dimensional in nature with 20,531 features versus 3,255 samples, we only used the top 2000 representative gene features selected from our mutual information analysis for final visualization.

Table 2. Top 11 gene mutual information to targeted cancer types values

GeneID	Breast	Lung	LungAD	Prostate	Kidney	Thyroid
A1BG-1	0.1887	0.0157	0.2199	0.1649	0.0396	0.1602
A1BG-AS1-503538	0.0253	0.0973	0.0973	0.1500	0.1322	0.0987
A2M-2	0.4592	0.8242	0.5156	0.0017	0.0296	0.0781
A4GALT-53947	0.0442	0.2646	0.0714	0.1426	0.1403	0.0559
AAAS-8086	0.1090	0.4013	0.2019	0.1209	0.1369	0.1075
AACS-65985	0.0000	0.3789	0.2312	0.0516	0.2836	0.3091
AADAT-51166	0.1901	0.1931	0.1843	0.1529	0.2833	0.1420
AAED1-195827	0.3013	0.3362	0.2249	0.0251	0.2003	0.2840
AAGAB-79719	0.2013	0.4644	0.2926	0.0750	0.5392	0.1456
AAK1-22848	0.0434	0.0444	0.0504	0.0033	0.3198	0.2729
AAMDC-28971	0.0746	0.3616	0.2100	0.0000	0.0000	0.1252

Following this, we employed a PCA (Principal Component Analysis) to reduce the dimension down to 50 features. Finally, utilizing t-Distributed Stochastic Neighbors Embedding (t-SNE) [13] we projected this down to two-dimensional space. We note that the first 50 components of the PCA were enough to account for 99% of the variance within the selected features.

t-SNE is a two-part dimensionality reduction scheme. First, t-SNE constructs a probability distribution over pairs of high-dimensional objects in such a way that similar objects have a high probability of being picked while dissimilar points have an extremely small probability of being picked. t-SNE then defines a similar probability distribution over the points in the low-dimensional map, minimizing the Kullback–Leibler divergence [14] between the two distributions with respect to the locations of the points in the map. As such, the model excels for visualization of high dimensional data.

PCA and t-SNE were done in conjunction for projection was because t-SNE does not scale well from an efficiency perspective across a large number of features and samples. Thus PCA was used initially to create set of features that would be both faithful to the original data and be appropriately sized for projection to two dimensions with t-SNE. t-SNE was used for the two dimensional visualization endpoint as it is desirable for the visualization to identify and cluster neighbors in an unsupervised and representative manner. The overall dimension reduction pipeline for unsupervised learning is depicted in Fig. 1.

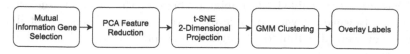

Fig. 1. Dimension reduction pipeline for unsupervised learning and visualization

Figure 2 depicts an initial visualization from our projection. The projection demonstrates a clear separation of the data across tissue origins. This indicates the importance of tissue origin in terms of significance of the features. In other words, the significant variation of RNA expression across samples with differing progeny is demonstrated clearly by the projection.

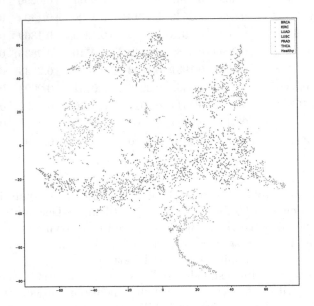

Fig. 2. Projection of gene expression of all 6 cancers and non-cancer data into two-dimensional feature space as per dimension reduction pipeline.

To further identify the distinguishability of origin from gene expression data, we utilized clustering analysis on the combined dataset. Our desired goal is to find out whether clustering will separate the samples based on the tissue type and status. In our analysis, we tentatively chose the cluster size as seven (non-cancer plus the six cancer types). We elected to use the Gaussian mixture model (GMM) for this clustering task [15]. We decided to use the Gaussian Mixture modeling because it is relatively efficient, we had a sufficient number of samples, and we had a known preliminary cluster number (7) based off of the types of cancer to base our analysis off of. In addition, Gaussian Mixture modeling is a relatively unbiased approach. Figure 3 shows the results of applying clustering calculation by displaying the contour lines of the GMM. We noticed that the model was successfully able to disambiguate between the various tissue origins of the cancers. However, it was not able to converge on a non-cancer distribution, because they were distributed throughout the space on the edge of the tissue origin clusters.

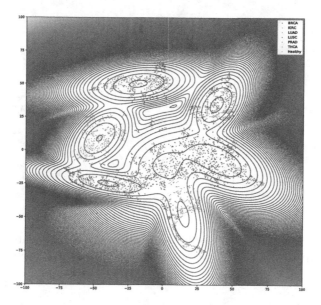

Fig. 3. Gaussian Mixture Model contour map superimposed on projection of gene expression of all 6 cancers and non-cancer data into two-dimensional feature space as per dimension reduction pipeline.

Because the prior visualization has healthy tissue aggregated into one category and is thus spread over the entire space, in order to determine the relative separability of status across sample tissue type we perform a further analysis for individual cancer types. Applying the same GMM modeling scheme to each individual cancer dataset. Setting the number of clusters to two (one for cancer, one for non-cancer), for each cancer type, we utilized the same feature reduction pipeline but only to data for that specific cancer type and tissue (including healthy data).

Figure 4 displays the two-dimensional data projection superimposed on the Gaussian mixture contour maps. From the following figures, we noticed that the cancer and healthy tissues are separable, but these samples are still relatively close in the projected spaces. With visible sample separation between cancer and healthy samples in the projected space for Breast and Kidney tissue, the use of an unsupervised machine learning model to predict the presence of cancer and tissue type is justified.

However, we found the Gaussian mixture model often elected to make a separation of the data based on clusters internal to the cancer subset rather than between non-cancer and cancer data. This is indicative of multimodality in the cancer data, which is supported by research suggesting that some cancers, such as Kidney cancer are in fact a collections of several different diseases of the kidney caused by several germline mutations which differentiate the various diseases. [16] Nevertheless, this is indicative that cancer and non-cancer samples can be separated by unsupervised methods alone and shown promise for supervised methods.

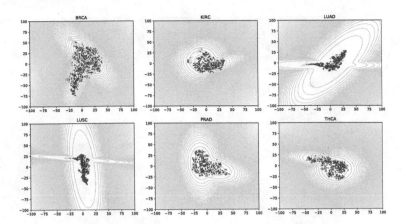

Fig. 4. Gaussian Mixture Model Results on dimensionally reduced data for each of the six cancers (purple) with their respective non-cancer samples (yellow). (Color Figure Online)

6   Predictive Model Generation

The results of the unsupervised learning techniques provided us the confidence and hope that we could develop a pan-cancer diagnostic model using the selected gene features. However, we still wish to determine whether a model using a limited subset of all gene expressions would be viable. In general, we should expect a model using a limited number of genes to be potentially more cost effective. As such, our next step was to explore such possibilities. Figure 5 illustrates the process of model generation and evaluation.

For each individual tissue type, we first a performed an 80-20 test train split on all six tissue sample datasets separately. The test train splits were performed separately in order to ensure both cancer and non-cancer data for each type of cancer is available and in an unbiased distribution in the final testing set. The individual training and testing sets were then combined, resulting in a mix of all six cancers as well as both cancer and non-cancer tissues samples within both sets. All non-cancer samples were aggregated into one class rather than being separate for each tissue type.

To test whether we could select a limited set of significant genes we first created a baseline set of top genes using the top 500 values from the mutual information. We then trained an RFE (Recursive Feature Elimination) model to select only 200 genes from the original 2,000 gene dataset. The RFE takes all 2,000 genes and selects the top genes first. Then, the removed genes were taken and placed back into the RFE model recursively, until only 200 genes remained. We compare these the results of training based on these two datasets to determine whether we have selected a limited set of genes without significant performance deterioration.

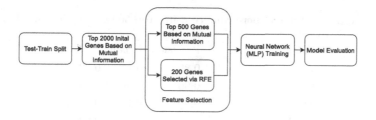

Fig. 5. Process of predictive model generation and validation

In both models, we created an MLP (Multilayer Perceptron) using the deep learning library Keras with five fully connected layers using the tanh activation function except for the final layer utilizing softmax. The model was trained using the Adam (Adaptive Moment Estimation) optimizer, an extension to the Stochastic Gradient Descent (SGD) optimizer, and the categorical cross-entropy loss function [17].

We trained the MLP across both datasets on the training data for each dataset. The training was stopped after 20 epochs following a reduction in the change in loss per epoch, and thus preventing overfitting to the data. Precision and recall are reported to quantify overall model and performance.

7   Results of Supervised Learning

Due to the imbalance between cancer and non-cancer samples within the dataset, we report precision and recall and F1 scores for each cancer. Results for the neural network trained on the 500 gene dataset with features selected via Mutual Information are presented in Table 3. Likewise, results for the 200 gene dataset with features selected via RFE are presented in Table 4. We note that the two models perform quite similarly despite using 300 less genes in the 200 gene model.

We also calculated the precision, recall and F1 score for each of the individual RFE models in their ability to analyze healthy data. The results of which are displayed in Table 5.

We report the multiclass precision, recall, and F1 Score for the 500 gene model in Table 6. These values are presented in terms of the micro, macro and weighted averages. The macro-average computes each metric independently for each class and then take the average, which does not account for class imbalance. On the other hand, the a micro-average aggregates the contributions of all classes in computing the average. Finally, the weighted average is similar to the macro average, but where each metric is weighted by the number of positive samples for each label.

Finally, we report the micro, macro, and weighted averages for precision, recall, and F1 Score for the RFE Model for each type of examined cancer in the RFE model in Table 7.

Table 3. MLP model performance using Top 500 genes without RFE

Cancer type	Precision	Recall	F1 Score
Breast	0.89	0.97	0.93
Kidney	0.96	0.99	0.75
Prostate	0.90	0.79	0.84
Thyroid	0.89	1.00	0.94
Lung Adenocarcinoma	0.95	0.96	0.67
Lung Squamous Cell	0.90	1.00	0.95

Table 4. MLP model performance using 200 genes with RFE

Cancer	Precision	Recall	F1 Score
Breast	0.96	0.98	0.97
Kidney	0.95	1.00	0.97
Prostate	0.97	0.97	0.97
Thyroid	0.84	0.98	0.9
Lung Adenocarcinoma	0.92	0.95	0.94
Lung Squamous Cell	0.97	0.92	0.94

Table 5. Healthy tissue prediction results

Tissue type	Precision	Recall	F1 Score
RFE Model			
Breast	0.25	0.07	0.11
Kidney	0.90	0.64	0.75
Thyroid	0.	0.	0.
Prostate	0.18	0.36	0.24
Lung Squamous Cell	1.00	0.14	0.25
Lung Adenocarcinoma	0.69	0.64	0.67
Mutual Information Model			
Combined	0.92	0.65	0.76

Table 6. 500 gene mutual information model averaged results

Average type	Precision	Recall	F1 Score
Micro	0.94	0.94	0.94
Macro	0.93	0.92	0.92
Weighted	0.94	0.94	0.93

Table 7. RFE model averaged results

RFE lung AD averages				RFE breast averages			
Average type	Precision	Recall	F1 Score	Average type	Precision	Recall	F1 Score
Micro	0.93	0.93	0.93	Micro	0.87	0.87	0.87
Macro	0.82	0.80	0.81	Macro	0.57	0.52	0.52
Weighted	0.92	0.93	0.93	Weighted	0.82	0.87	0.84
RFE thyroid averages				RFE kidney averages			
Average type	Precision	Recall	F1 Score	Average Type	Precision	Recall	F1 Score
Micro	0.89	0.89	0.89	Micro	0.95	0.95	0.95
Macro	0.44	0.50	0.47	Macro	0.93	0.82	0.86
Weighted	0.78	0.89	0.83	Weighted	0.95	0.95	0.95
RFE prostate averages				RFE lung SQ averages			
Average type	Precision	Recall	F1 score	Average type	Precision	Recall	F1 score
Micro	0.74	0.74	0.74	Micro	0.90	0.90	0.90
Macro	0.54	0.57	0.54	Macro	0.95	0.57	0.60
Weighted	0.82	0.74	0.77	Weighted	0.91	0.90	0.87

8   Discussion and Conclusion

In this research, we applied both unsupervised and supervised approaches to confirm that there is a significant difference in gene expressions from extracted tissues of varying origins, we created and evaluated neural network models for analysis and classification of tissue type and status from tissue sample gene expression levels, and we analyzed the advantages and drawbacks of the usage of both the 500 and 200 gene set for model training.

The results of the t-SNE and Gaussian Mixture Model analysis displayed clear separation between clusters of cancer samples. As such, the high accuracy of the supervised machine learning models is to be expected. Overall, the best performing model was the top 500 MI model, however, the RFE 200 models performed similarly in cancer classification while using less than half the number of genes.

Overall, the top 500 gene MI model performed the best in terms of F1 scores, achieving an F1 score exceeding 0.76, but the 200 gene RFE model demonstrated strong results in the classification of both lung cancers and kidney cancer. This indicated that fewer genes could be used in diagnosis with only a small trade off in model performance, which would enable more cost-effective sequencing.

Our research indicates that the use of machine learning models to differentiate between different cancers is possible. We hypothesized that with the increased depth provided by RNA sequencing data, a machine learning model trained on TCGA RNA-sequencing data for cross-cancer classification would be possible. We produced a Gaussian Mixture Model, which in conjunction with t-SNE and PCA dimensionality reduction, demonstrated the separability of gene expression data across tissue type and status. We then created a neural network capable of taking genetic data for 500 genes from an unknown sample and predicting the tissue of origin. Finally, we analyzed the effectiveness of using an RFE model to select relevant genes in reducing the needed genes for classification. The models created in this research can serve as the groundwork for future cross-cancer studies and provides evidence that cross-cancer models are viable in medical diagnoses.

Overdiagnosis in breast and prostate cancer, along with misdiagnosis found in lung cancer demonstrate the need for a model to differentiate between cancers and to accurately predict the health of a patient in question. The use of mutual information in our models is a novel approach to the problem of a cross-cancer diagnosis. However, this work was limited by a large class imbalance found within the data. Despite initial success, this method provides only the basic framework for future work in cross-cancer analysis. The accuracies and F1 scores achieved are still below that which is necessary for medical diagnosis.

In the future, we would also like to explore other feature selection criteria. For example, we could create a logistic regression with LASSO to select the top 500 genes from the top 2,000 mutual information list, in a similar line to our approach with RFE. We also intend to test the viability of introducing healthy data from other sources such as the GTEx project [18] in order to increase the number of healthy samples found within the data to improve the poor accuracies found in healthy tissue classification in all the models.

References

1. Siegel, R.L., Miller, K.D., Jemal, A.: Cancer statistics, 2019. CA: Cancer J. Clin. **69**(1), 7–34 (2019)
2. Draisma, G., et al.: Lead time and overdiagnosis in prostate-specific antigen screening: importance of methods and context. JNCI J. Natl. Cancer Inst. **101**(6), 374–383 (2009)
3. Serefoglu, E.C., Altinova, S., Ugras, N.S., Akincioglu, E., Asil, E., Balbay, D.: How reliable is 12-core prostate biopsy procedure in the detection of prostate cancer? Can. Urol. Assoc. J. **6**(2) (2012)
4. Bhatt, M., Kant, S., Bhaskar, R.: Pulmonary tuberculosis as differential diagnosis of lung cancer. South Asian J. Cancer **1**(1), 36 (2012)
5. Rogler, C.E.: RNA expression microarrays (REMs), a high-throughput method to measure differences in gene expression in diverse biological samples. Nucleic Acids Res. **32**(15), e120–e120 (2004)
6. Ozsolak, F., Milos, P.M.: RNA sequencing: advances, challenges and opportunities. Nat. Rev. Genet. **12**(2), 87–98 (2011)
7. Zhao, S., Fung-Leung, W.P., Bittner, A., Ngo, K., Liu, X.: Comparison of RNA-Seq and microarray in transcriptome profiling of activated T cells. PLoS ONE **9**(1), e78644 (2014)
8. The Cancer Genome Atlas Research Network, et al.: The Cancer Genome Atlas Pan-Cancer analysis project. Nat. Genet. **45**(10), 1113–1120 (2013)
9. Gao, K., Wang, D., Huang, Y.: Cross-cancer prediction: a novel machine learning approach to discover molecular targets for development of treatments for multiple cancers. Cancer Inform. **17**, 117693511880539 (2018)

10. Golcuk, G., Tuncel, M.A., Canakoglu, A.: Exploiting ladder networks for gene expression classification. In: Rojas, I., Ortuño, F. (eds.) IWBBIO 2018. LNCS, vol. 10813, pp. 270–278. Springer, Cham (2018). https://doi.org/10.1007/978-3-319-78723-7_23

11. Casey, M., Chen, B., Zhou, J., Zhou, N.: A machine learning approach to prostate cancer risk classification through use of RNA sequencing data. In: Chen, K., Seshadri, S., Zhang, L.-J. (eds.) BIGDATA 2019. LNCS, vol. 11514, pp. 65–79. Springer, Cham (2019). https://doi.org/10.1007/978-3-030-23551-2_5

12. Wei, L., Jin, Z., Yang, S., Xu, Y., Zhu, Y., Ji, Y.: TCGA-assembler 2: software pipeline for retrieval and processing of TCGA/CPTAC data. Bioinformatics **34**(9), 1615–1617 (2018)

13. van der Maaten, L., Hinton, G.: Visualizing data using t-SNE. J. Mach. Learn. Res. **9**, 2579–2605 (2008)

14. Kullback, S.: Information Theory and Statistics. Reprint edn. Smith, Gloucester, Mass (1978) OCLC: 187308462

15. Lindsay, B.G.: Mixture models: Theory, geometry and applications. NSF-CBMS Regional Conference Series in Probability and Statistics **5**, i–163 (1995)

16. Linehan, W.M.: Genetic basis of kidney cancer: role of genomics for the development of disease-based therapeutics. Genome Res. **22**(11), 2089–2100 (2012)

17. Kingma, D.P., Ba, J.: Adam: A Method for Stochastic Optimization. arXiv:1412.6980 [cs] (January 2017)

18. Consortium, G.: The Genotype-Tissue Expression (GTEx) project. Nat. Genet. **45**(6), 580–585 (2013)

19. Bolton, K.L.: Association between BRCA1 and BRCA2 mutations and survival in women with invasive epithelial ovarian cancer. JAMA **307**(4), 382 (2012)

20. Amin, M.B., et al. (eds.): AJCC Cancer Staging Manual, Eight edn. American Joint Committee on Cancer, Springer, Chicago (2017)

21. Network, C.G.A.: Genomic classification of cutaneous melanoma. Cell **161**(7), 1681–1696 (2015)

22. Lee, J., Kim, D.W.: Feature selection for multi-label classification using multivariate mutual information. Pattern Recogn. Lett. **34**(3), 349–357 (2013)

23. Peng, H., Long, F., Ding, C.: Feature selection based on mutual information criteria of max-dependency, max-relevance, and min-redundancy. IEEE Trans. Pattern Anal. Mach. Intell. **27**(8), 1226–1238 (2005)

24. Nadon, R., Shoemaker, J.: Statistical issues with microarrays: processing and analysis. Trends Genet. **18**(5), 265–271 (2002)

25. Welch, H.G., Prorok, P.C., O'Malley, A.J., Kramer, B.S.: Breast-cancer tumor size, overdiagnosis, and mammography screening effectiveness. New England J. Med. **375**(15), 1438–1447 (2016)

26. Cronin, K.A., et al.: Annual report to the nation on the status of cancer, part i: national cancer statistics: annual report national cancer statistics. Cancer **124**(13), 2785–2800 (2018)

Heavy Vehicle Classification Through Deep Learning

Pei-Yun Sun$^{(\boxtimes)}$, Wan-Yun Sun, Yicheng Jin, and Richard O. Sinnott

School of Computing and Information Systems, The University of Melbourne,
Melbourne, VIC, Australia
{pssun, wanyun, rsinnott}@unimelb.edu.au,
yichengj@student.unimelb.edu.au

Abstract. Understanding the flow of traffic on road networks is increasingly important especially with the continued urbanization of the global population. Numerous hardware and software technologies have been applied to measure traffic volumes by Government agencies and/or organizations such as Google, however they are either expensive to deploy; limited in their ability to disambiguate the kinds of vehicles on the road network, or of increasing importance, they infringe on the privacy of individuals, e.g. tracking phones. In this paper we describe work applying deep learning technologies to identify and classify different vehicles on the road network of Victoria with specific focus on heavy goods vehicles (trucks and trailers). Specifically, we present an approach to automatically detect, classify and count the unique classes of trucks and trailers that are found on the road network and the direction of travel. We apply and compare leading deep learning approaches including You Only Look Once version 3 (YOLOv3) and Single Shot Multi-Box Detector (SSD). This paper builds upon earlier work [1] which focused on data (video) from a single traffic junction in Melbourne. This work is based on a wider range of data (videos) from locations reflecting the diversity of road use including multi-lane motorways, rural roads and city roads.

Keywords: Trucks · Trailers · Deep learning · SSD · YOLOv3 · Deep SORT

1 Introduction

The global population is increasingly living in cities. This is the case in Melbourne which continues to increase in population year on year. Indeed, it is expected that 50% of the total population of Australia will live in two cities in the coming decade: Melbourne and Sydney. Whilst new houses and schools can be built to accommodate the population growth, one major challenge that exists is the road network and traffic issues. The road network has not changed significantly in many years and narrow roads designed for historic traffic patterns are simply not suitable for the population growth. One core component of traffic flows is heavy goods vehicles such as trucks (lorries) and trailers. There are many forms of trucks and trailers, and the regulations that apply to each kind can vary, e.g. certain trucks may be prohibited from using certain routes or if a diversion occurs then certain routes may be unsuitable for certain trucks. At present

© Springer Nature Switzerland AG 2020
S. Nepal et al. (Eds.): BIGDATA 2020, LNCS 12402, pp. 220–236, 2020.
https://doi.org/10.1007/978-3-030-59612-5_16

there is no system to identify the types of trucks and trailers that use the road network. Overhead video cameras exist on many parts of the road network of Victoria however these are used to stream data for manual inspection of the traffic flows at the head-quarters of the main transport agency in Victoria, VicRoads (www.vicroads.vic.gov.au). It is highly desirable to automate the identification and classification of the different types of trucks and trailers on the road network. This is the focus of this paper. In particular, we explore two leading deep learning approaches for image detection and classification: You Only Look Once version 3 (YOLOv3) and Single Shot Detection (SSD). We apply these models to diverse scenarios reflecting the real-world complexity of traffic flows in Melbourne.

The rest following part of the paper is structured as follows. In Sect. 2, we describe the truck and trailer standard classification types and data sets from diverse locations that were used to train the deep learning models. In Sect. 3, we compare the performance of YOLOv3 and SSD on the data. In Sect. 4, we analyse the results of YOLOv3 on different sites with unique characteristics/challenges for heavy goods vehicle classification. In Sect. 5, we present a method for identifying the direction of travel of vehicles. In Sect. 6 we present related work and finally in Sect. 7, we conclude the work and identify potential areas for future work.

2 Heavy Goods Dataset

One of the challenges for deep learning is establishment of high-quality data used for training. Such data should contain many images with diverse features related to the objects of interest with appropriate and accurate labelling of these images. In our work reported in [1], a 43-h video file captured by a surveillance camera at Francis Street (Melbourne) was provided (by VicRoads). This location is near to the port of Melbourne and has many types of trucks and trailers that travel through in both daytime and night-time. For data collection, each heavy vehicle in the video required individual frames to be selected and the image of the truck and trailer manually added to an appropriately labelled image dataset using the annotation tool *LabelImg* [2]. According to the National Heavy Vehicle Regulator of Australia (NHVR - www.nhvr.gov.au) heavy goods vehicles can be divided into different categories: *Rigid Trucks; Semi-trailer Combinations; Rigid Truck and Trailer Combinations; B-Double Combinations; Common Type 1 Road Trains,* and *Common Type 2 Road Trains.* The outline of these trucks and trailers are shown in Fig. 1. These trucks and trailers have different lengths and weights (both when loaded and unloaded) that can restrict the roads upon which they are allowed to drive. Within these general classes, the trucks in each class can be further divided into more specific classes as shown in Fig. 1. This is typically based on the number of axels and the spacing between the axels.

In addition to these finer-grained and general classes of heavy goods vehicles, several new classes have been introduced from the work described in [1] including Dangerous Goods Vehicles (DGV), buses and vans. A DGV can be identified by a small diamond sign on the vehicle. Based on the work described in [1] which was based on a single, slow moving junction, the work presented here deals with more realistic traffic issues reflecting the diverse traffic flows in Melbourne.

National Heavy Vehicle Regulator
Common Heavy Freight
Vehicle Configurations (NHVR)

	Description	Maximum Length (metres)	Maximum Regulatory Mass under GML (tonnes)	Maximum Regulatory Mass under CML (tonnes)	Maximum Regulatory Mass under HML (tonnes)
1. COMMON RIGID TRUCKS - GENERAL ACCESS					
(a)	2 Axle Rigid Truck	≤ 12.5	15.0	CML does not apply	-
(b)	3 Axle Rigid Truck	≤ 12.5	22.5	23.0	-
(c)	4 Axle Rigid Truck	≤ 12.5	26.0	27.0	-
(d)	4 Axle Twinsteer Rigid Truck	≤ 12.5	26.5	27.0	-
(e)	5 Axle Twinsteer Rigid Truck	≤ 12.5	30.0	31.0	-
2. COMMON SEMITRAILER COMBINATIONS - GENERAL ACCESS					
(a)	3 Axle Semitrailer	≤ 19.0	24.0	-	-
(b)	4 Axle Semitrailer	≤ 19.0	31.5	32.0	32.0
(c)	5 Axle Semitrailer	≤ 19.0	35.0	36.0	37.5
(d)	5 Axle Semitrailer	≤ 19.0	39.0	40.0	40.0
(e)	6 Axle Semitrailer	≤ 19.0	42.5	43.5	45.5
3. COMMON RIGID TRUCK AND TRAILER COMBINATIONS (General access when complying with prescribed mass and dimension requirements)					
(a)	2 Axle Truck and 2 Axle Dog Trailer	≤ 19.0	30.0	-	-
(b)	2 Axle Truck and 2 Axle Pig Trailer	≤ 19.0	30.0	CML does not apply	-
(c)	3 Axle Truck and 2 Axle Dog Trailer	≤ 19.0	40.5	41.0	-
(d)	3 Axle Truck and 2 Axle Pig Trailer	≤ 19.0	37.5	CML does not apply	-
(e)	3 Axle Truck and 3 Axle Dog Trailer	≤ 19.0	42.5	43.5	-
(f)	3 Axle Truck and 3 Axle Pig Trailer	≤ 19.0	40.5	CML does not apply	-
(g)	3 Axle Truck and 4 Axle Dog Trailer	≤ 19.0	42.5	43.5	-
(h)	4 Axle Truck and 3 Axle Dog Trailer	≤ 19.0	42.5	43.5	-
(i)	4 Axle Truck and 4 Axle Dog Trailer	≤ 19.0	42.5	43.5	-
4. COMMON B-DOUBLE COMBINATIONS - CLASS 2					
(a)	7 Axle B-double	≤ 19.0	55.5	57.0	57.0
(b)	8 Axle B-double	≤ 26.0	59.0	61.0	62.5
(c)	8 Axle B-double	≤ 26.0	59.0	61.0	62.5
(d)	9 Axle B-double	≤ 26.0	62.5	64.5	68.0
5. COMMON TYPE 1 ROAD TRAINS - CLASS 2					
(a)	9 Axle A-double	≤ 36.5	72.0	74.0	74.0
(b)	11 Axle A-double	≤ 36.5	79.0	81.0	85.0
(c)	12 Axle A-double	≤ 36.5	82.5	84.5	90.5
(d)	12 Axle Modular B-triple	≤ 35.0	82.5	84.5	90.5
(e)	12 Axle B-triple	≤ 36.5	82.5	84.5	90.5
(f)	14 Axle AB-triple	≤ 36.5	99.0	101.0	107.5
(g)	15 Axle AB-triple	≤ 36.5	102.5	104.5	113.0
(h)	11 Axle Rigid Truck and 2 Dog Trailers	≤ 36.5	88.5	90.5	91.0

Fig. 1. Truck and trailer classification

Table 1. Site location, nature of location type, video duration and number of objects classified

Site - location	Nature of location	Video	#Objects
Site 00 - *Francis St*	Slow moving 16-lane junctions	41 h	9,905
Site 01- *Stanley Ave*	8-lane bidirectional Motorway	24 h	30,413
Site 02 - *Somerville Rd*	Industrial dual carriageway	24 h	5,635
Site 03 - *Moreland & Napier St*	Complex urban junctions	24 h	8,770
Site 04 - *St Kilda Rd*	Bidirectional dual carriageway	48 h	2,257
Site 05 -*Woolsthorpe Heywood Rd*	Rural Road	24 h	105

Table 2 shows the manual classification and count of vehicles that formed the basis for the training data set. There were 8 general classes and 38 finer-grained classes.

These data sets were split into the ratio: 8:1:1 for training, testing and validation respectively. It is worth noting that test data set did not contain all classes since the

Fig. 2. Example views of data capture at Sites 00–05

Table 2. Distribution of heavy goods vehicle based on manual classification

General class	Finer class	Count
Rigid Truck	2 Axle Rigid Truck	8,880
	3 Axle Rigid Truck	5,392
	4 Axle Rigid Truck	25
	4 Axle Twinsteer Rigid Truck	1,396
	5 Axle Twinsteer Rigid Truck	162
Semitrailer	3 Axle Semitrailer	692
	4 Axle Semitrailer	285
	5 Axle Semitrailer	949
	6 Axle Semitrailer	10,548
	7 Axle Semitrailer	91

(continued)

Table 2. (*continued*)

General class	Finer class	Count
Truck and Trailer	2 Axle Truck And 2 Axle Dog Trailer	8
	2 Axle Truck And 2 Axle Pig Trailer	138
	3 Axle Truck And 2 Axle Dog Trailer	15
	3 Axle Truck And 2 Axle Pig Trailer	79
	3 Axle Truck And 3 Axle Dog Trailer	471
	3 Axle Truck And 3 Axle Pig Trailer	14
	3 Axle Truck And 4 Axle Dog Trailer	378
	3 Axle Truck And 5 Axle Dog Trailer	55
	3 Axle Truck And 6 Axle Dog Trailer	79
	4 Axle Truck And 2 Axle Dog Trailer	1
	4 Axle Truck And 2 Axle Pig Trailer	5
	4 Axle Truck And 3 Axle Dog Trailer	13
	4 Axle Truck And 4 Axle Dog Trailer	10
4-11 Axle B Double	4 Axle B-double	13
	5 Axle B-double	32
	6 Axle B-double	103
	7 Axle B-double	453
	8 Axle B-double	308
	9 Axle B-double	2,085
	10 Axle B-double	62
	11 Axle B-double	114
8-11 Axle A Double	8 Axle A-double	2
	9 Axle A-double	43
	11 Axle A-double	240
Van or Bus	Van	13,065
	Bus	835
DGV (signs)	DGV (signs)	2,537
Other	Other	670
Total		**50,248**

number of some classes was small. The data set was also unbalanced with some vehicles appearing rarely and others being very common. Based on these experiments, it was decided to remove classes with less than 10 counts in the final test set.

3 YOLOv3 and SSD Classification Comparison

In this section, we focus on one-stage deep learning detection methods: YOLOv3 [3] and SSD [4]. The structure and configuration of both YOLOv3 and SSD and the mean average precision (mAP) used are described in detail in [1]. In this paper, we applied a new version of YOLOv3 [5], which optimizes the connection structure of the network and provides a more powerful way to extract features by adding several max pooling layers to obtain information of different scales.

3.1 Aggregated Truck and Trailer Detection with SSD

Figure 3 (left) shows the mAP for individual classes of vehicle for all sites listed in Table 1. This shows a mAP of 37.29%, with a maximum AP of 75% for the *9 Axle B-double*. The reason for such good detection for the *9 Axle B-double* is that it has unique features that can be used to distinguish it from other classes. There are also many such images in the training data set (Table 2). SSD has difficulties however when the targets are small, e.g. DGV signs cannot be seen, nor remote vehicles (far lanes) on the motorway.

Figure 3 (right) shows the mAP for the general class case where it is seen that the mAP increases to 56.54%. *Semitrailer, 4–11 Axle B-double* and *Rigid Truck* can be identified with reasonable accuracy, however the others perform less well. This is likely due to the *8–11 Axle A-double* having insufficient data for the model to learn enough features for the classification.

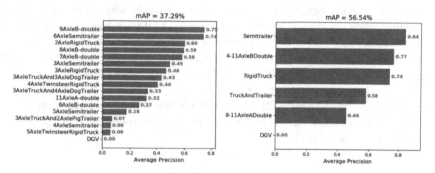

Fig. 3. mAP for finer-grained (left) and general (right) classification for all sites using SSD

Unsurprisingly, the mAP of the general class case is higher than that for classifying finer grained classes. It is easier for the SSD model to make correct decisions with fewer classes and more data. Nevertheless, a large number of erroneous classifications happen, e.g. a *3Axle Rigid Truck* has a high probability to be erroneously identified as *2 Axle Rigid Truck*, however they both belong to the general class *Rigid Truck*.

3.2 Aggregated Truck and Trailer Detection with YOLOv3

Figure 4 (left) shows the mAP (76.86%) for finer-grained classification across all sites using YOLOv3. Almost every class has a higher AP than SSD. A *6Axle Semitrailer* has the highest mAP at 91%. This vehicle has specific features and a large number of data samples. However, the mAP for DGV is low, but considering the small size of DGV signs (approx. 25 cm), this is a promising result. Figure 3 (right) shows the mAP for the general classification across all sites, where the mAP reaches 79.09%. Comparing Fig. 4 (left) and Fig. 4 (right), unlike SSD, the mAP for general classification by YOLOv3 is only marginally higher than that for the finer grained classification. This shows that YOLOv3 is more stable than SSD.

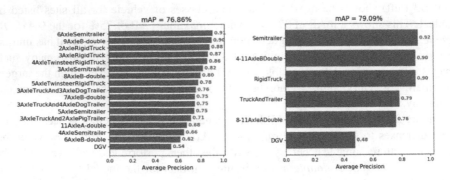

Fig. 4. mAP for finer-grained (left) and general (right) classification for all sites using YOLOv3

YOLOv3 is clearly better than SSD for detecting trucks and trailers. It uses multi-scale feature maps to detect objects. In addition, YOLOv3 uses a more powerful network (darknet-53), while the network for SSD is based on VGG-16. From an accuracy perspective, the mAP of YOLOv3 is 39.57% higher than SSD for finer-grained classification, and 22.55% higher for general classification. Importantly, YOLOv3 can detect small objects, e.g. DGV signs on trucks, and traffic on distant motorway lanes. The one downside of YOLOv3 is that it needs slightly more memory than SSD to run: YOLOv3 needs at least 1.4 G memory to detect an image or frame, while SSD only needs 1.0 G.

4   Individual Site Comparisons

One major contribution of this work compared to [1] is dealing with diverse data and heterogeneous settings as illustrated in Fig. 2. These comprise slow moving multi-directional crossroads; bidirectional dual carriageways; fast moving motorways with multiple near/far lanes as well as video from rural settings including fast moving vehicles. In this section, we consider the performance of YOLOv3 for the general and finer-grained classification for each of the six different sites and the challenges they raise.

As mentioned in Sect. 2, classes which have a ground truth lower than ten in testing are excluded for the final evaluation. This exclusion was not applied to Site05 as the ground truth for all classes was below ten objects in total, i.e. there were minimal trucks and trailers on the rural road. Table 4 and Table 5 show the performances of the general and individual truck and trailer detection for each of the six sites respectively. The Yolov3 model is efficient and works reliably in each of the sites with mAP ranging from 0.73 to 1 for the general class detection, and from 0.77 to 0.93 for the individual truck and trailer vehicle classification. The model thus provides an acceptable degree of genericity and robustness across different sites and the challenges they give rise to (Table 3).

Table 3. General truck and trailer classification accuracy for each Site (Yolov3)

	Site00	Site01	Site02	Site03	Site04	Site05
Overall mAP	0.87	0.73	0.91	0.77	0.97	1
4-11AxleBDouble	0.96	0.81	0.95	0.91	–	1
Semitrailer	0.96	0.92	0.94	0.85	–	1
RigidTruck	0.92	0.89	0.94	0.86	0.97	1
TruckAndTrailer	0.94	0.77	0.81	–	–	1
DGV	0.55	0.15	–	0.45	–	–
8-11AxleADouble	–	0.85	–	–	–	–

Table 4. Individual truck and trailer classification accuracy for each Site (Yolov3)

	Site00	Site01	Site02	Site03	Site04	Site05
Overall mAP	0.86	0.77	0.86	0.80	0.97	0.93
9AxleB-double	0.99	0.81	1	0.92	–	1
7AxleB-double	0.93	–	–	–	–	–
6AxleSemitrailer	0.96	0.9	0.93	0.85	–	1
2AxleRigidTruck	0.92	0.87	0.93	0.83	0.97	1
5AxleTwinsteerRigidTruck	–	–	–	0.89	–	–
3AxleRigidTruck	0.85	0.86	0.94	0.82	–	–
4AxleTwinsteerRigidTruck	0.86	0.91	0.83	0.84	–	–
8AxleB-double	0.86	–	–	–	–	–
4AxleSemitrailer	–	0.86	–	–	–	–
3AxleSemitrailer	0.69	–	0.89	0.92	–	–
3AxleTruckAnd2AxlePigTrailer	–	0.82	–	–	–	–
3AxleTruckAnd4AxleDogTrailer	0.9	0.72	–	–	–	0.67
5AxleSemitrailer	0.91	0.7	0.48	0.66	–	1
3AxleTruckAnd3AxleDogTrailer	–	0.74	–	–	–	–
11AxleA-double	–	0.73	–	–	–	–
DGV	0.62	0.27	–	0.51	–	–

As seen, Site04 and Site05 have the best detection results, whilst Site00 and Site02 have superior results to Site01 and Site03. Site04 and Site05 both achieved a mAP of over 0.97 for the general classification, and over 0.93 for the individual classification. Site00 and Site02 reached a mAP of approximately 0.86 for the general classification and above 0.87 for the individual classification.

As seen in Fig. 2, the video captured by the cameras has different challenges that can impact results: an increase in complexity of road geometry and occlusion issues, the distance between camera and vehicles (e.g. remote vehicles in site01), and acuteness of camera angle (site05) can make classification non-trivial, e.g. compared to the slow moving truck shown in Fig. 2 (site00). Sites with the best results (Site02, Site04 and Site05) have a simple road geometry with few lanes and minimize occlusion (hidden or partially visible vehicles). Although Site00 has a complex road geometry it is a crossroad with slow moving vehicles and camera angle that provides an adequate perspective to see the vehicles. On the other hand, Site03 is a forked road with a low camera angle while Site01 had eight lanes in total with traffic moving at over 100 km/h. Nevertheless, the results and generality of the solution for all settings is encouraging.

5   Vehicle Direction Through Deep SORT

In addition to detecting the trucks and trailers it is also necessary to track the direction of travel of vehicles in real time. To tackle this, the work applied Multiple Object Tracking (MOT) algorithms focused on Detection Based Tracking. Specifically, we considered two MOT algorithms: SORT [6], and Deep SORT [7]. Kalman filters [8] and the Hungarian algorithm [9] are two techniques based on the SORT algorithm. Kalman filters are used where the object of interest is in the next frame, whilst the Hungarian algorithm is used for associating a detected bounding box where the object of interest will be in the next frame.

Target Representation. In SORT, the following representation is used to capture the state of each target object. In this representation, the position of the target center is denoted by (u, v), and the scale and aspect ratio of the target's bounding box is represented by (s, r). The remaining three components express the respective velocities: (u, v, s). In this algorithm, the aspect ratio r is assumed to be constant and therefore, it does not have a corresponding velocity component.

$$x = [u, v, s, r, \dot{u}, \dot{v}, \dot{s}] \tag{1}$$

Predict and Update Target State. When tracking objects, the SORT algorithm runs two steps iteratively: *predict* and *update*. In the predict step, SORT predicts the future state of each target (in the next frame), while in the update step, SORT updates the target state if there is a detection bounding box associated with the predicted bounding box of the target. By running these two steps constantly, it is possible to link bounding boxes across video frames to capture the trajectories of targets.

Bounding Box Assignment. In the update step, the detection boxes of the current frame need to be assigned to the predicted bounding boxes. The assignments are decided by the Intersection over Union (*IOU*) between the detection bounding boxes and the predicted bounding boxes. To tackle this, it is necessary to construct an assignment matrix with IOU distance $(1 - IOU)$ as the cost. As it is a one-to-one assignment problem, this can be solved using the Hungarian algorithm. During the optimization, each detection box will only be assigned to a predicted box if the *IOU* between them is greater than a user-defined threshold: IOU_{min}. This is only assigned if the boxes are close enough. In SORT, the tracking algorithm depends on the position and velocity of the objects. If multiple objects share a similar state, e.g. position and velocity, ID switches can happen. To solve this, Wojke proposed an extension to SORT - Deep SORT [7]. This algorithm incorporates appearance features using a pre-trained convolutional neural network. With the new appearance features, it is possible to reduce the number of ID switches and increase the robustness, e.g. in case of occlusions.

Target Representation. The target representation used in Deep SORT is almost the same as in SORT, except it uses an aspect ratio and height (g, h) and velocities $(\dot{g}, \dot{h})$, while SORT uses a scale and aspect ratio (s, r) and velocity components:

$$\mathrm{x} = \begin{bmatrix} \mathrm{u}, \mathrm{v}, \mathrm{g}, \mathrm{h}, \dot{\mathrm{u}}, \dot{\mathrm{v}}, \dot{\mathrm{g}}, \dot{\mathrm{h}} \end{bmatrix} \qquad (2)$$

In Deep SORT, Kalman filters are used to make predictions and update states in the same way as SORT.

Deep Appearance Descriptor. The network for computing appearance features in this work was pre-trained on a person re-identification dataset called Motion Analysis and Re-identification Set (MARS) [10]. This dataset contains more than 1.1 million images of 1,261 pedestrians hence it is large enough to train a CNN model for discriminating pedestrians. As shown in Fig. 5, the output of this model is a normalized appearance feature vector of length 128.

Name	Patch Size/Stride	Output Size
Conv 1	$3 \times 3/1$	$32 \times 128 \times 64$
Conv 2	$3 \times 3/1$	$32 \times 128 \times 64$
Max Pool 3	$3 \times 3/2$	$32 \times 64 \times 32$
Residual 4	$3 \times 3/1$	$32 \times 64 \times 32$
Residual 5	$3 \times 3/1$	$32 \times 64 \times 32$
Residual 6	$3 \times 3/2$	$64 \times 32 \times 16$
Residual 7	$3 \times 3/1$	$64 \times 32 \times 16$
Residual 8	$3 \times 3/2$	$128 \times 16 \times 8$
Residual 9	$3 \times 3/1$	$128 \times 16 \times 8$
Dense 10		128
Batch and ℓ_2 normalization		128

Listing 1 Matching Cascade

Input: Track indices $\mathcal{T} = \{1, \dots, N\}$, Detection indices $\mathcal{D} = \{1, \dots, M\}$, Maximum age $A_{\max}$
1: Compute cost matrix $C = [c_{i,j}]$ using Eq. 5
2: Compute gate matrix $B = [b_{i,j}]$ using Eq. 6
3: Initialize set of matches $\mathcal{M} \leftarrow \emptyset$
4: Initialize set of unmatched detections $\mathcal{U} \leftarrow \mathcal{D}$
5: **for** $n \in \{1, \dots, A_{\max}\}$ **do**
6:    Select tracks by age $\mathcal{T}_n \leftarrow \{i \in \mathcal{T} \mid a_i = n\}$
7:    $[x_{i,j}] \leftarrow \min\text{-}cost\text{-}matching(C, \mathcal{T}_n, \mathcal{U})$
8:    $\mathcal{M} \leftarrow \mathcal{M} \cup \{(i,j) \mid b_{i,j} \cdot x_{i,j} > 0\}$
9:    $\mathcal{U} \leftarrow \mathcal{U} \setminus \{j \mid \sum_i b_{i,j} \cdot x_{i,j} > 0\}$
10: **end for**
11: **return** $\mathcal{M}, \mathcal{U}$

Fig. 5. Deep appearance descriptor architecture (left) and Matching Cascade (right) used in Deep SORT [7]

New Assignment Metric. In SORT, the costs in the assignment matrix are based on the IOU distance between each detection and prediction box. Here, the IOU distance cost is replaced with a new metric that combines motion and appearance information by a weighted average using a hyperparameter λ to control the weights. The new cost between prediction i and detection j can be calculated as:

$$c_{i,j} = \lambda d^{(1)}(i,j) + (1 - \lambda)d^{(2)}(i,j) \tag{3}$$

In the formula above, $d^{(1)}(i,j)$ measures the squared Mahalanobis distance between the predicted boxes y_i and the detected box d_j, where both boxes are represented by four values: (u, v, g, h). This distance also takes uncertainties of predictions into account by including standard deviation S_i of track i, for when larger uncertainties for the predictions exhibiting longer periods of occlusions arise. In other words, the distance, $d^{(1)}(i,j)$ measures how many standard deviations the detected box is away from the predicted box of the object trajectory.

$$d^{(1)}(i,j) = (d_j - y_i)^T S_i^{-1} (d_j - y_i) \tag{4}$$

The second part of the cost formula involves $d^{(2)}(i,j)$, which measures the distance (dissimilarity) between track i and detection j in terms of their appearance based on 128-dimensional feature values computed by the CNN model. The value of $d^{(2)}(i,j)$ is given by the minimum cosine distance between each appearance descriptor r_k in the gallery of track i and descriptor r_j of detection j. The gallery R of each track stores up to 100 appearance descriptors of the bounding boxes assigned to the track as recorded in the previous frames.

$$d^{(2)}(i,j) = \min\{1 - r_j^T r_k^{(i)} | r_k^{(i)} \in R_i\} \tag{5}$$

Matching Cascade. After defining the costs for the assignment matrix, Deep SORT uses a matching cascade to weaken the bias in the Mahalanobis distance. During the calculation of assignment costs, the Mahalanobis distance has a bias towards predictions with larger uncertainty, since the predictions with larger standard deviations tend to have smaller $d^{(1)}(i,j)$ with the detection box. The cascade solves this problem by assigning bounding boxes to the *younger age* tracks first, where the age is defined as the number of frames since the first measurement (detection). When assigning the detections to tracks, the assignments are valid only if they are within a gating region formed by two assignment metrics (i.e. $b_{i,j} = 1$):

$$b_{i,j} = \Pi_{m=1}^2 b_{i,j}^{(m)} \tag{6}$$

Each gate $b^{(m)}$ indicates whether the distance is below an upper limit $t^{(m)}$:

$$b_{i,j}^{(m)} = 1\left[d^{(m)}(i,j) \le t^{(m)}\right] \tag{7}$$

In Sect. 3 and 4, we identified that YOLOv3 was more suitable for detecting trucks and trailers. Therefore, it was chosen to be the basis for the object detection model. The outputs of YOLOv3 were thus fed into the Deep SORT algorithm for tracking vehicles.

Confidence Threshold. As the MOT algorithms are based on the vehicle detection results, it is necessary to improve the detection model. The simplest way is to find a good confidence threshold for outputs if the confidence of detection is below a given threshold t_{conf}, if so, the detected bounding box is ignored in that frame. Using a high confidence threshold could however give increased false negatives, i.e. correct detections with low confidence may be ignored. While using a low confidence threshold may lead to false positives, i.e. wrong detections being made. By tuning the threshold, it was found that setting $t_{conf} = 0.4$ was a suitable value to minimize errors.

Deep SORT Parameter Setting. The final parameters used for Deep SORT are shown in Table 6. In addition to the parameters discussed in the previous sections, n_{init} is used to indicate the number of consecutive detections before a track is confirmed. In the implementation of Deep SORT, if there is any missing detection before a track is confirmed, the track will be deleted immediately. However, as the detection model is not perfect, setting n_{init} to any value other than 1 will make the algorithm keep assigning new ID's to vehicles until the detection model successfully detects it in n_{init} number of consecutive frames. Therefore, the parameter n_{init} was set to 1.

Table 5. Final parameter for Deep SORT

Parameter	Value	Description
t_conf	0.4	Min confidence threshold for detection
t_{nms}	0.5	Non-maximal suppression IOU threshold
max$\_age$	30	Max age of tracker (=Matching cascade depth)
$max\_mahalanobis$	9.4877	Gating threshold t_1: Mahalanobis distance for boxes
$max\_cos\_distance$	0.4	Gating threshold t_2: cosine distance for appearance
n_{init}	1	Num of consecutive detections before track is confirmed

Once we obtain vehicle tracks, we use them to identify the direction of travel of trucks and trailers. The approach we used is to pre-define *In-Out* regions comprising North, South, East, and West (see Fig. 6), and compare pre-defined regions with the start and end points of the track, where the start and end points of the tracks are defined as the centres of the first and last detection box of each track. For example, if the start point of track lies in North-In (NI) region, and its end point lies in East-Out (EO) region. Then, the travelling direction of this track is North-East (NE). This method allows the model to identify the direction of travel for any site however it requires the user to provide the in- and out- regions in advance. That is, determining the orientation of the cameras needs to be established at some point.

To demonstrate the travelling direction, we randomly picked a 5-min video from Site 00 to test the proposed method. In this sample video, 25 trucks and trailers passed by. 76% of them had both start and end points correctly identified; 80% of them had the start point correctly identified and 88% of them had the end point correctly identified. From Table 7, we see that the method is able to obtain the correct direction even if there are id switches. Among the 6 incorrect identifications of directions, 4 of them were related to the South or East lane, either identifying South as West or vice versa. This is because the in- and out- regions are very similar, especially the in- regions, since the trucks and trailers from South and East almost enter the video from the same region. The remaining two incorrect identifications have N/A for the direction. The first is because the *4AxleB-double* is moving too fast and hence its track doesn't have enough detections to ensure it is a valid track. The second N/A is caused by occlusion, i.e. there is another *6Axle Semitrailer* in front, so it missed the start point within the in-region.

Fig. 6. Pre-defined in- and out- regions of Site 00 (left) and Site 01 (right)

Table 6. Results of identifying direction of travel (Site 00)

Time	Type	Detection	Ground truth	ID switches	Correct
00:48	6AxleSemitrailer	WS	WS	Y	Y
00:58	6AxleSemitrailer	WS	WS	Y	Y
01:03	8AxleB-double	WS	WS	N	Y
01:10	2AxleRigidTruck	WN	WN	N	Y
01:17	9AxleB-double	NE	NS	N	N
01:28	9AxleB-double	NS	NS	N	Y
01:38	4AxleB-double	N/A	NS	N	N
01:53	6AxleSemitrailer	WS	WS	N	Y
01:58	6AxleSemitailer	WS	WS	N	Y
02:03	6AxleSemitrailer	WS	WS	N	Y

(continued)

Table 6. (*continued*)

Time	Type	Detection	Ground truth	ID switches	Correct
02:05	5AxleSemitrailer	N/A	WE	N	N
02:08	2AxleRigidTruck	WS	WS	N	Y
02:28	2AxleRigidTruck	SW	EW	N	N
02:58	9AxleB-double	NS	NS	N	Y
03:15	6AxleSemitrailer	SN	SN	Y	Y
03:46	3AxleRigidTruck	WS	WS	Y	Y
03:51	5AxleTwinsteerRigidTruck	WS	WS	N	Y
03:54	2AxleRigidTruck	WS	WS	Y	Y
03:56	6AxleSemitrailer	WE	WE	N	Y
04:04	3AxleRigidTruck	WE	WE	N	Y
04:08	6AxleSemitrailer	SW	EW	Y	N
04:17	6AxleSemitrailer	WE	WE	N	Y
04:20	6AxleSemitrailer	SW	EW	Y	N

6 Related Work

Since its release in 2018, YOLOv3 has been widely applied for object recognition and detections in diverse areas such as traffic signs recognition [11], pedestrian detection [12], and cholelithiasis identification and gallstone classification [13]. Improvements and variations of the framework are continually emerging, e.g. an improved YOLOv3 network based on DenseNet [14] to deal with low resolution of feature layers, and a Gaussian YOLOv3 [15] to cope with mis-localization during autonomous driving.

Multiple object tracking is increasingly demand in various fields. This technique has been used for pose-tracking in augmented reality [16], multi-camera visual surveillance [17], autonomous driving [18], UAV reconnaissance [19], and driver face tracking [20]. Researchers are continually developing new algorithms with better performance and new datasets for evaluations.

Imaging datasets have been established for different purposes. TrackingNet [21] focuses on object tracking in the wild; MOT20 is designed for pedestrian tracking; Need-for-Speed (NFS) [22] consists of high frame rate (240 FPS) videos captured from comprehensive real-world scenarios. UA-DETRAC [23] was established for traffic monitoring. Other researchers have improved existing MOT algorithms like Deep SORT for improved performance. Hou improved the Deep SORT algorithm by incorporating a lower confidence filter [24]. Bertinetto used a novel fully convolutional Siamese networks for object tracking [25]. Fan improved the network by combining it with region proposal network (RPN) [26] while Zhang improved the performance using a deeper and wider backbone for Siamese networks [27].

7 Conclusions

In this paper we trained and evaluated YOLOv3 and SSD models for trucks and trailer detection using datasets obtained from videos across six different sites around Melbourne. These sites had unique challenges for vehicle classification reflecting real world scenarios. We considered the mean average precision (mAP) of general and individual class models. By comparing the results of YOLOv3 models and SSD models, we identified that YOLOv3 performed significantly better than SSD in detection, with a mAP of 76.86% for the general model and 79.09% for individual classifications.

We further evaluated the results of applying YOLOv3 across different sites and found that both the general and individual YOLOv3 model had sufficient genericity to efficiently perform detection across the various sites. The complexity of the road geometry, the distance between camera and vehicles, the speed of traffic flow, and camera angles all have an impact on the overall performance.

Finally, we considered the direction of travel using Deep SORT and used of In- and Out- regions reflecting vehicle trajectories. We achieved an overall accuracy of 76%.

There are many possible extensions to this work. Inclusion of other (all vehicles) and determining the vehicle speed are two immediate extensions to the work. More practically, deploying the models across the road cameras across Victoria is a further follow on that is currently being considered.

The authors would like to thank the Roads Cooperation of Victoria (VicRoads) and especially Ben Phillips and Ben Atkinson for providing the video files used to generate the training dataset and the initial idea for the work that has been realised here. The authors also acknowledge use of the combined HPC/GPU facility (SPARTAN) at the University of Melbourne. SPARTAN was part-funded by an Australian Research Council LIEF grant.

References

1. Chen, L., Sun, P.-Y., Jia, Y., Sinnott, R.O.: Identification and classification of trucks and trailers on the road network through deep learning. In: BDCAT 2019 - Proceedings of the 6th IEEE/ACM International Conference on Big Data Computing, Applications and Technologies (2019). https://doi.org/10.1145/3365109.3368781
2. Tzutalin: tzutalin/labelImg (2015). https://github.com/tzutalin/labelImg
3. Redmon, J., Farhadi, A.: Yolov3: an incremental improvement. arXiv preprint arXiv:1804.02767 (2018)
4. Liu, W., et al.: SSD: single shot multibox detector. In: Leibe, B., Matas, J., Sebe, N., Welling, M. (eds.) ECCV 2016. LNCS, vol. 9905, pp. 21–37. Springer, Cham (2016). https://doi.org/10.1007/978-3-319-46448-0_2
5. Huang, Z., Wang, J.: DC-SPP-YOLO: dense connection and spatial pyramid pooling based YOLO for object detection. arXiv preprint arXiv:1903.08589 (2019)
6. Bewley, A., Ge, Z., Ott, L., Ramos, F., Upcroft, B.: Simple online and realtime tracking. In: 2016 IEEE International Conference on Image Processing (ICIP), pp. 3464–3468 (2016)

7. Wojke, N., Bewley, A., Paulus, D.: Simple online and realtime tracking with a deep association metric. In: 2017 IEEE International Conference on Image Processing (ICIP), pp. 3645–3649 (2017)
8. Kalman, R.E.: A new approach to linear filtering and prediction problems. J. Basic Eng. **82**, 35–45 (1960)
9. Kuhn, H.W.: The Hungarian method for the assignment problem. Naval Res. Logistics Q. **2**, 83–97 (1955)
10. Zheng, L., et al.: MARS: a video benchmark for large-scale person re-identification. In: Leibe, B., Matas, J., Sebe, N., Welling, M. (eds.) ECCV 2016. LNCS, vol. 9910, pp. 868–884. Springer, Cham (2016). https://doi.org/10.1007/978-3-319-46466-4_52
11. Rajendran, S.P., Shine, L., Pradeep, R., Vijayaraghavan, S.: Real-time traffic sign recognition using YOLOv3 based detector. In: 2019 10th International Conference on Computing, Communication and Networking Technologies (ICCCNT), pp. 1–7 (2019)
12. Valiati, G.R., Menotti, D.: Detecting pedestrians with YOLOv3 and semantic segmentation infusion. In: 2019 International Conference on Systems, Signals and Image Processing (IWSSIP), pp. 95–100 (2019)
13. Pang, S., et al.: A novel YOLOv3-arch model for identifying cholelithiasis and classifying gallstones on CT images. PloS One **14**, e0217647 (2019)
14. Tian, Y., Yang, G., Wang, Z., Wang, H., Li, E., Liang, Z.: Apple detection during different growth stages in orchards using the improved YOLO-V3 model. Comput. Electron. Agric. **157**, 417–426 (2019)
15. Choi, J., Chun, D., Kim, H., Lee, H.-J.: Gaussian yolov3: an accurate and fast object detector using localization uncertainty for autonomous driving. In: Proceedings of the IEEE International Conference on Computer Vision, pp. 502–511 (2019)
16. Uchiyama, H., Marchand, E.: Object detection and pose tracking for augmented reality: recent approaches. Presented at the (2012)
17. Wang, X.: Intelligent multi-camera video surveillance: a review. Pattern Recogn. Lett. **34**, 3–19 (2013)
18. Darms, M., Rybski, P., Urmson, C.: Classification and tracking of dynamic objects with multiple sensors for autonomous driving in urban environments. In: 2008 IEEE Intelligent Vehicles Symposium, pp. 1197–1202 (2008)
19. Kapania, S., Saini, D., Goyal, S., Thakur, N., Jain, R., Nagrath, P.: Multi object tracking with UAVs using deep SORT and YOLOv3 RetinaNet detection framework. In: Proceedings of the 1st ACM Workshop on Autonomous and Intelligent Mobile Systems. Association for Computing Machinery, New York (2020). https://doi.org/10.1145/3377283.3377284
20. Wang, Q., Cao, L., Xia, J., Zhang, Y., et al.: MTCNN-KCF-deepSort: driver face detection and tracking algorithm based on cascaded kernel correlation filtering and deep-SORT (2020)
21. Muller, M., Bibi, A., Giancola, S., Alsubaihi, S., Ghanem, B.: Trackingnet: a large-scale dataset and benchmark for object tracking in the wild. In: Proceedings of the European Conference on Computer Vision (ECCV), pp. 300–317 (2018)
22. Galoogahi, H.K., Fagg, A., Huang, C., Ramanan, D., Lucey, S.: Need for speed: a benchmark for higher frame rate object tracking. arXiv preprint arXiv:1703.05884 (2017)
23. Wen, L., et al.: UA-DETRAC: a new benchmark and protocol for multi-object detection and tracking. Comput. Vis. Image Underst. **193**, 102907 (2020)
24. Hou, X., Wang, Y., Chau, L.-P.: Vehicle tracking using deep SORT with low confidence track filtering. In: 2019 16th IEEE International Conference on Advanced Video and Signal Based Surveillance (AVSS), pp. 1–6 (2019)
25. Bertinetto, L., Valmadre, J., Henriques, J.F., Vedaldi, A., Torr, P.H.S.: Fully-convolutional siamese networks for object tracking. In: Hua, G., Jégou, H. (eds.) ECCV 2016. LNCS, vol. 9914, pp. 850–865. Springer, Cham (2016). https://doi.org/10.1007/978-3-319-48881-3_56

26. Fan, H., Ling, H.: Siamese cascaded region proposal networks for real-time visual tracking. In: Proceedings of the IEEE Conference on Computer Vision and Pattern Recognition, pp. 7952–7961 (2019)
27. Zhang, Z., Peng, H.: Deeper and wider siamese networks for real-time visual tracking. In: Proceedings of the IEEE Conference on Computer Vision and Pattern Recognition, pp. 4591–4600 (2019)

Short Paper Track

Spatial Association Pattern Mining Using In-Memory Computational Framework

Jin Soung Yoo[(✉)], Wentao Shao, and Kanika Binzani

Purdue University, Fort Wayne, IN 46805, USA
{yooj,shaow01,binzk01}@pfw.edu

Abstract. Spatial association pattern mining is a useful spatial data mining task for discovering interesting relationship patterns of spatial features based on spatial proximity. Spatial data mining is known as data-intensive computing. The explosive growth of spatial data demands computationally efficient methods for analyzing large complex data. Parallel and distributed computing is effective for large-scale data mining algorithms. This paper presents spatial association pattern mining with Spark which is a specially-designed in-memory parallel computing platform. The performance of Spark-based method proposed is compared with its corresponding MapReduce-based method.

1 Introduction

As the advance of location sensing, mobile computing and storage technology, rich spatial and spatio-temporal data is being generated from numerous data sources including mobile phones, social media, GPS tracking systems, wireless sensors, and outbreaks of disease. The spatial and spatio-temporal data is considered nuggets of valuable information [14]. Spatial data mining [12] is a computational process to discover interesting and previously unknown, but potentially useful patterns from large spatial or spatio-temporal databases. As one of spatial data mining tasks, spatial association pattern mining [9] is used for discovering interesting relationship patterns among spatial features and possibly some non-spatial features. Spatial association mining is useful in many application domains [10,15,17] including public health, social science, environmental science, urban planning and business. An example of spatial association patterns is uncovered relationship patterns between certain deceases (e.g., cholera) and nearby environmental risk factors (e.g., contaminated water reservoirs). Spatio-temporal association pattern mining can be used for understanding spatial epidemic dynamics of disease outbreak in an area.

The explosive growth of georeferenced data has emphasized the need to develop computationally efficient methods for data mining tasks. A massive amount of data available these days is far beyond the capacity of a single machine. Due to the initial memory capacity and computational capability of a single machine, traditional single node-based algorithms suffer from large scale

© Springer Nature Switzerland AG 2020
S. Nepal et al. (Eds.): BIGDATA 2020, LNCS 12402, pp. 239–246, 2020.
https://doi.org/10.1007/978-3-030-59612-5_17

data processing. In order to discover spatial association patterns, frequent feature sets must first be found from an input dataset with mining task parameters such as frequent threshold and neighbor distance threshold. Memory requirement for handling all candidate feature sets and pattern instances can blow up fast with the increase of data size. When spatial queries for pattern instance search are embedded in the mining process, the computation is more complex and expensive.

Modern computing platforms facilitating the distributed execution of massive tasks are becoming increasingly popular. A MapReduce-like system such as Hadoop [2] has been proven to be an efficient framework for big data analysis for many applications, e.g., machine learning and graph processing. Although Hadoop has been developed as a cluster computing system for handling large-scale data processing, but the MapReduce system does not meet the performance expectation of many iterative algorithms for machine learning and data mining because it does a complete disk read/write for each iteration. On the other hand, Spark [3] is alternative cluster computing infrastructure which is a specially-designed in-memory parallel computing framework. It is highly suitable for iterative algorithms, and supports batch, interactive and stream processing of data. This work presents a spatial association pattern mining algorithm with Spark and compares its performance with a MapReduce-based method.

The remainder of this paper is organized as follows. Section 2 describes our spatial association pattern mining problem. Section 3 introduces cloud computing platforms. Section 4 describes a parallel spatial association mining method with Spark. In Sect. 5, the experimental result is presented, and this paper concludes in Sect. 6.

2   Spatial Association Pattern Mining

Spatial and spatio-temporal association mining problems have been popularly studied since the problem of mining association rules based on spatial relationships (e.g., proximity, adjacency, etc.) was first presented by Koperski and Han [9]. Most of works on mining spatial association patterns use one of two representative pattern models; reference-based model [9] and co-location model [13]. Our work follows the reference-based spatial association model which is useful for discovering spatial features frequently associated with a specific focal feature, for example, a crime type or a disease outbreak type. A spatial association pattern is often presented with spatial and non-spatial predicates, in the form of $P_1 \wedge \ldots \wedge P_m \to Q_1 \wedge \ldots \wedge Q_n(s,c)$ where at least one of the predicates $P_1,\ldots,P_m, Q_1,\ldots,Q_n$ is a spatial predicate. For example, $is\_a(x, robbery) \wedge occur(x, 1pm - 3pm) \wedge within(x, zip\ code\ 46805\ area) \to close\_to(x, school)$, where x represents an object of a focal feature, e.g., a robbery incident. A spatial predicate is used to describe a spatial relationship between the focal feature and other features, e.g., $close\_to(x, school)$. A non-spatial predicate describes the focal feature object's property, e.g., $occur(x, 3pm - 5pm)$.

Our problem of spatial association pattern mining can be statements as the followings:

Given:

1) A focal feature type f and its instance object set D_f where $o_i \in D_f$ is presented with a vector $< focal\_feature\_type, instance\_id, latitude, longitude,$ $other\ property\ attributes>$.

2) A set of task-relevant spatial features $F = \{f_1, \ldots, f_n\}$ where $f_i \neq f$, the focal feature, $i = 1, \ldots, n$ and a set of their instance objects $D = D_1 \cup D_2, \ldots, \cup D_n$ where $o \in D$ is simply presented with a vector $< feature\_type, instance\_id, latitude, longitude>$.

3) A spatial neighbor relationship R, i.e., a distance function and a neighbor distance threshold (min_dist).

4) A minimum frequent threshold (min_freq) and a minimum confidence threshold (min_conf).

Find: Spatial association patterns.

We present the neighboring information of each focal feature object with a transaction record format and call it a *neighborhood record*. For each object $o_i \in D_f$ of the focal feature f, the neighborhood record n_i is defined to $\{f_j : f_j \neq f \land o_j \in D \land distance(o_i, o_j) \leq min\_dist\}$. The neighborhood records of all the focal feature objects, $N = \{n_1, n_2, \ldots, n_m\}$ are generated, where $m = |D_f|$. We then combine the property features $a_i = \{attr_1, \ldots attr_h\}$ of each focal object o_i with its neighborhood record n_i, and simply call $t_i = \{n_i \cup a_i\}$ the *transaction record* of the focal object o_i. We can find frequent association patterns from the transaction record set $T = \{t_1, t_2, \ldots, t_m\}$ using a frequent itemset mining algorithm such as Apriori [5]. We use the *support* metric for measuring the frequency of spatial association patterns. The support s of a feature set I is defined as $\frac{|t_i : I \subset t_i \in T|}{|T|}$. If the support value s is greater than a minimum support value min_freq, the set I becomes a set of spatial features which are often observed in the neighborhood areas of the focal feature f and its attribute features. The frequent feature set I is used to generate rules associated with f such as $f \cup I_i \to f \cup I_j$, where $I_i \cup I_j = I$ and $I_i \cap I_j = \emptyset$. The confidence c of a rule $f \cup I_i \to f \cup I_j$ is defined as $\frac{|t_p : I_i \cup I_j \subset t_p \in T|}{|t_q : I_i \subset t_q \in T|}$.

3   Cloud Computing Platforms

Modern cloud computing platforms such as Hadoop [2] and Spark [3] have been popularly used for large-scale data processing and various applications including machine learning and graph processing. MapReduce [8] is a programming model for expressing distributed computations on massive amounts of data and an execution framework for large-scale data processing on clusters of commodity servers. The MapReduce system was originally developed by Google and has since enjoyed widespread adoption via an open-source implementation called Hadoop. MapReduce simplifies parallel processing by abstracting away the complexities involved in working with distributed systems, such as computational

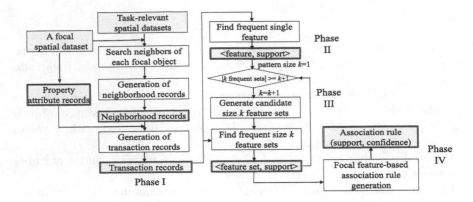

Fig. 1. Spatial association pattern mining work flow

parallelization, work distribution, and dealing with unreliable hardware and software. A MapReduce job is executed in two main phases of user defined data transformation functions, namely, *map* and *reduce*.

Spark [18] is another distributed computing framework and offers a number of features to make big data processing fast. The fundamental feature is its in-memory parallel execution model in which all data will be loaded into memory. The second major feature is that Spark can provide very flexible DAG-based (directed acyclic graph) data flow. These characteristics can significantly speedup the computation of iterative algorithms such as Apriori algorithm [5] and some of machine learning algorithms. The programming model of Spark is upon a new distributed memory abstraction called *resilient distributed datasets* (RDDs) which is proposed for in-memory computation in a large cluster. A RDD is an immutable collection of data records. Spark provides a variety of built-in operations to transform one RDD to another RDD. RDDs can also achieve fault-tolerance based on lineage information rather than replication. There are several co-location mining works with MapReduce or Graphics Processing Units (GPUs) [6,7,11,16]. However it is hard to find reference-based spatial association mining works with Spark.

4   Parallel and Distributed Spatial Association Mining with Spark

A work flow for mining spatial association patterns from a focal feature dataset and task relevant feature datasets is presented in Fig. 1. Phase I generates the transaction record of each focal feature object from input datasets. Phase II finds frequent singletons from the transaction records. A Spark transform function, $flatMap()$ is applied on each of records to get all features from the transaction records. Then a $mapToPair()$ function is applied to transform all feature items to the key/value pairs of <feature, 1>. A $reduceByKey()$ function is invoked to count the frequency of each of features, and then using a $filter()$ function, we

```
Input: Neighborhood transaction RDD with frequent singletons, RDD_d,
       Frequent feature sets in iteration 2, L_2
       A minimum frequent threshold θ
Output: Frequent size k(≥ 2) feature sets L_k
Variables: RDD_c: instance sets of candidates,
           RDD_p: (feature instance set, 1),
           RDD_r: (feature set, support),
           RDD_f: (frequent feature set, support),
1) k=2
2) while (|L_{k-1}| ≥ k) do
3)    C_k ← candidateGeneration(L_{k-1})
4)    broadcast(C_k)
5)    RDD_c ← RDD_d.flatMap(C_k)
6)    RDD_p ← RDD_c.mapToPair()
7)    RDD_r ← RDD_p.groupByKey()
8)    RDD_f ← RDD_r.filter(θ)
9)    RDD_f.saveAsTextFile("size k output")
10)   L_k ← RDD_f.collect()
11)   k ← k + 1
12)do while
```

Algorithm 1: Size k frequent feature set mining with Spark(Phase-III)

eliminate infrequent features whose frequency is less than the support threshold. Remaining features become frequent singletons.

Phase III finds size k (≥ 2) frequent features. Algorithm 1 shows the pseudo code of Phase III. Candidate sets of size k are generated having all possible combinations of size $k-1$ whose every possible nonempty subsets are frequent. For example, after all frequent feature sets of size 2 are discovered, these feature sets are used to create the candidate sets of size 3. Then we scan the transaction record RDD (or transaction records filtered with frequent singletons) to list all instances of each candidate feature sets by applying a $flatMap()$ function. Further we generate <feature set, 1> key/value pairs for each instance of candidate feature sets. Finally the $reduceByKey()$ function collects all the support counts of each of candidates. We find frequent size k feature sets whose count is greater than or equal to the minimum support threshold using a $filter()$ function. This procedure iterates until the number of size $k-1$ frequent feature sets discovered is less than k. The process of finding frequent feature sets takes a majority of the computation time of association rule mining. Hash Tree and Trie are often considered for the underlying data structure. Phase IV is the last step of generating association rules from frequent feature subsets.

5   Experiment

We conducted the experimental evaluation of the proposed Spark-based method with comparing its performance with a corresponding MapReduce-based method. Algorithm 2 shows the pseudo code of the MapReduce-based algorithm. Both programs are implemented using Java and Spark or MapReduce APIs.

For the experiment, we used Amazon EMR platforms [1]. The EMR computing platform provides resizable computing capability in the cloud. We used small clusters of general purpose machines for master node and worker nodes. Each core had 16 GiB RAM and 4 virtual CPUs. We used Spark Clusters with Spark 2.4.4 on Hadoop 2.8.5 YARN, and also used Hadoop Clusters with Hadoop 2.8.5. All input datasets and output files were stored in Amazon S3 storage.

For this experiment, we used real-world data as well as synthesis data. We collected crime incident records from a Police Department website [4]. The criminal incident record has six attributes such as incident no, date, time, nature, address and community. The 'nature' attribute describes a crime type like theft and burglary. For this experiment, we focused on theft. For the task-relevant data, points of interest (POI) in the study area were used. The dataset includes 767 data points with 20 features such as bank, mall, park and restaurant. Each POI record is described with its latitude, longitude, name, category, address and other description attributes. We used only latitude, longitude, and category which serves as feature type. In addition, we prepared two synthetic datasets of focal data points: D1(5K) and D2(20K). A task relevant synthetic dataset was also generated with 50K points with 200 features.

We conducted the experimental evaluation with changing the number of focal feature data points, the neighborhood size and the minimum frequent threshold. Figure 2 (a) shows the relative performance of the Spark-based method to the MapReduce-based method with the real-world datasets on base clusters with one master node and two worker nodes. The minimum support was fixed to 0.1, and the neighbor distance threshold was changed in each experiment. The Spark-based method was faster approximately 85% than the MapReduce-based method under this experimental setting. The performance difference was larger when the mining task had many iterative phases. In the second experiment, we increased the number of worker nodes. In the experiment with the real-world datasets (Fig. 2 (b)), we could not find noticeable difference with the increase

Mapper (key=o_i, value=$neighTransRecord$)
```
1) setup() {
2)    L_{k-1} ← readCacheFile()
3)    C_k ← candidateGeneration(L_{k-1})
4) }
5) featureSets ← findInstanceSets(neighTransRecord, C_k)
6) foreach set ∈ featureSets do
7)   emit(set, 1);
8) end do
```

Reducer (key=$feature\ set$, value=$[1]$)
```
1) support ← sum([1]);
2) if support ≥ θ then
3)   emit(feature set, support);
4) end if
```

Algorithm 2: MapReduce-based method

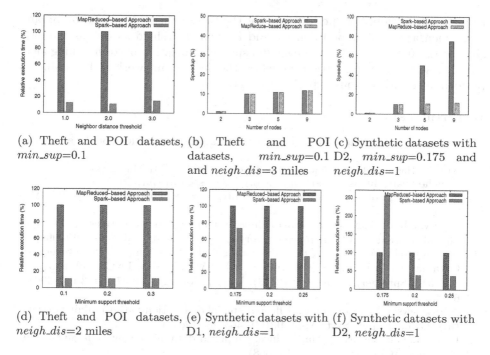

(a) Theft and POI datasets, min_sup=0.1

(b) Theft and POI datasets, min_sup=0.1 and $neigh\_dis$=3 miles

(c) Synthetic datasets with D2, min_sup=0.175 and $neigh\_dis$=1

(d) Theft and POI datasets, $neigh\_dis$=2 miles

(e) Synthetic datasets with D1, $neigh\_dis$=1

(f) Synthetic datasets with D2, $neigh\_dis$=1

Fig. 2. Experiment results (a) by neighbor distance, (b)–(c) by number of cluster nodes, and (d)–(f) by frequent threshold

of nodes. However in the experiment with the synthetic datasets (Fig. 2 (c)), the Spark-based method showed the speedup of approximately 70%. In the third experiment, we examined the effect of the support threshold. In the experiment with the real-world data sets (Fig. 2 (d)), the Spark-based method takes approximately 90% less time than the MapReduce-based method. In the experiment with the synthetic datasets (Fig. 2 (e) and (f)), the Spark-based method showed overall better performance than the other method. However in the experiment setting with the D2 dataset and the support threshold 0.175, the Spark-based method's performance was suffered due to large candidate feature sets and memory limitation.

6   Conclusion

This paper presents a spatial association pattern mining algorithm with Spark, and compared its computational performance with the corresponding MapReduce-based method. The experiment result showed that the Spark-based approach achieves overall significant speedup than the MapReduce-based approach. The in-memory parallel execution model of Spark especially benefits the iterative computation for finding all frequent feature subsets. However, the Spark-based method started to slow down or fail with behaving badly due to resource starvation. The performance of the two methods depended on parameters such as neighborhood density and frequent threshold as well as the data size. In the future, we plan to develop a spatial association pattern mining algorithm based on the co-location model on the Spark framework and examine its performance with various experimental settings.

References

1. Amazon Elastic MapReduce. http://aws.amazon.com/elasticmapreduce/. Accessed 31 May 2020
2. Apache Hadoop. http://hadoop.apache.org/. Accessed 31 May 2020
3. Apache Spark. https://spark.apache.org/. Accessed 31 May 2020
4. Fort Wayne Indiana Police Department. http://www.fwpd.org/. Accessed 31 May 2020
5. Agarwal, R., Srikant, R.: Fast algorithms for mining association rules. In: Proceedings of International Conference on Very Large Databases, pp. 487–499 (1994)
6. Andrzejewski, W., Boinski, P.: GPU-accelerated collocation pattern discovery. In: Catania, B., Guerrini, G., Pokorný, J. (eds.) ADBIS 2013. LNCS, vol. 8133, pp. 302–315. Springer, Heidelberg (2013). https://doi.org/10.1007/978-3-642-40683-6_23
7. Andrzejewski, W., Boinski, P.: Efficient spatial co-location pattern mining on multiple GPUs. Expert Syst. Appl. 93(C), 465–483 (2018)
8. Dean, J., Ghemawat, S.: MapReduce: simplified data processing on large clusters. Commun. ACM 51(1), 107–113 (2008)
9. Koperski, K., Han, J.: Discovery of spatial association rules in geographic information databases. In: Proceedings of International Symposium on Large Spatial Data Bases, pp. 47–66 (1995)
10. Li, J., Adilmagambetov, A., Mohomed, S.M.J., Zaïane, O.R., Osornio-Vargas, A., Wine, O.: On discovering co-location patterns in datasets: a case study of pollutants and child cancers. Geoinformatica 20(4), 651–692 (2016)
11. Sainju, A.M., Jiang, Z.: Grid-based colocation mining algorithms on GPU for big spatial event data: a summary of results. In: Gertz, M., et al. (eds.) SSTD 2017. LNCS, vol. 10411, pp. 263–280. Springer, Cham (2017). https://doi.org/10.1007/978-3-319-64367-0_14
12. Shekhar, S., Chawla, S.: Spatial Databases: A Tour. Prentice Hall (2003)
13. Shekhar, S., Huang, Y.: Co-location rules mining: a summary of results. In: Proceedings of International Symposium on Spatio and Temporal Database (2001)
14. Vatsavai, R.R., Ganguly, A., Chandola, V., Stefanidis, A., Klasky, S., Shekhar, S.: Spatiotemporal data mining in the era of big spatial data: algorithms and applications. In: Proceedings of ACM SIGSPATIAL International Workshop on Analytics for Big Geospatial Data, pp. 1–10 (2012)
15. Weiler, M., Schmid, K.A., Mamoulis, N., Renz, M.: Geo-social co-location mining. In: Proceedings of International ACM Workshop on Managing and Mining Enriched Geo-Spatial Data, pp. 19–24 (2015)
16. Yoo, J.S., Doulware, B., Kimmey, D.: Parallel co-location mining with MapReduce and NoSQL systems. Knowl. Inf. Syst. 62, 1433–1463 (2020)
17. Yu, W.: Spatial co-location pattern mining for location-based services in road networks. Expert Syst. Appl. 46, 324–335 (2016)
18. Zaharia, M., et al.: Resilient distributed datasets: a fault-tolerant abstraction for in-memory cluster computing. In: Proceedings of the USENIX Conference on Networked Systems Design and Implementation, p. 2 (2012)

Dissecting Biological Functions for BRCA Genes and Their Targeting MicroRNAs Within Eight Clusters

Yining Zhu[1], Ethan Sun[2], and Yongsheng Bai[3(✉)]

[1] British International School of Houston, Houston, TX 77494, USA
[2] Seven Lakes High School, Katy, TX 77450, USA
[3] Next-Gen Intelligent Science Training, Ann Arbor, MI 48105, USA
bioinformaticsresearchtomorrow@gmail.com

Abstract. The Cancer Genome Atlas (TCGA) estimated that 12.4% of women born in the US will develop breast cancer in their lives. The goal of this study is to identify common biological signatures (ex. Gene Ontology or GO terms) for breast cancer candidate genes and miRNAs among these eight clusters identified through bioinformatics method in our recent study. The eight "communities" or clusters for Breast Invasive Carcinoma (BRCA) generated in our previous study are performed for functional annotation in this study. We observed that among all the GO terms enriched in top five groups of these eight clusters, Transcription, Lumen, Nucleolus, and Nucleoplasm are enriched the most, followed by Nucleoside binding, Nucleotide binding, and ATP binding. Two clusters among eight contain three or more previously reported breast cancer risk genes. We examined these clusters in terms of pathway association for miRNAs targeting the breast cancer risk genes, and found that 18 of 26 targeting miRNAs are involved in breast cancer related pathway PI3K-Akt. Our study showed that many miRNAs targeting breast cancer cluster genes are also associated with breast cancer related pathways, which provide evidence that some miRNA and gene pairs likely contribute to breast cancer in the context of their targeting relationship.

Keywords: BRCA · TCGA · MiRNA

1 Introduction

Breast cancer has been a very popular kind of cancer in women, placed in second (after skin cancer) in commonality. Based on the cancer statistics for the years 2007 through 2009, The Cancer Genome Atlas (TCGA) estimated that 12.4% of women born in the US will develop breast cancer in their lives (https://www.cancer.gov/types/breast). Although scientists have found out that mutations of BRCA1 and BRCA2 genes each account for about 10% of breast cancer and mutations in other 42 genes account for another 10%, causes of the remaining 70% are still unknown [1]. MicroRNAs (MiRNAs) are abundant non-coding RNAs that participate in post-transcriptional regulation through binding to the 3' UTRs of messenger RNAs (mRNAs). The analysis of gene and miRNA interaction is complex. In a previous study performed by Huang et al.

© Springer Nature Switzerland AG 2020
S. Nepal et al. (Eds.): BIGDATA 2020, LNCS 12402, pp. 247–251, 2020.
https://doi.org/10.1007/978-3-030-59612-5_18

2018 [2], the authors used 33 breast cancer signatures and corresponding gene lists to examine overlapping functions with METABRIC (Molecular Taxonomy of Breast Cancer International Consortium) data.

Clustering techniques are often used to classify groups with similar signatures. The genes that contribute to the same diseases or cancer are often grouped or clustered together. Existing approaches only analyzed the genetic signatures within each group of genes. The Louvain algorithm was compared with fast greedy, walktrap, leading_eigen, label_propagation, edge_betweenness, and MWMM [3] methods in identifying miRNA-mRNA "communities" or clusters within a bipartite graph, accounting for simultaneous comparison of tumor and normal samples. Most existing graph-based clustering algorithms consider the topology of one single instance (e.g., gene of miRNA) and treat all nodes equivalently in the graph, like MAGIA2 [4] and miRmapper [5]. Louvain algorithm [6] can cluster mRNA–miRNA interaction pairs into functional miRNA–mRNA regulatory modules. The goal of this study is to identify common biological signatures (ex. Gene Ontology or GO terms) for genes and MiRNAs among eight clusters that were identified in our previous study [7] to be associated with breast cancer.

2   Methods

2.1   Cluster Selection

The gene clusters we used in this study were taken from our previous study [7]. Specifically, a modified Louvain algorithm [6] was run to detect the "communities" or clusters on 20,661 miRNA-mRNA pairs with inverse correlation. We only considered cluster results for Breast Invasive Carcinoma (BRCA) to conduct analysis in this study. There were a total of eight clusters for BRCA performed for analysis in this study. The number of genes in each cluster is listed in Table 1.

2.2   DAVID Go Analysis

We selected all eight significant clusters for BRCA and annotated biological functions for cluster genes with the tool Database for Annotation, Visualization, and Integrated Discovery or DAVID [8, 9] using the "OFFICIAL_GENE_SYMBOL" identifier. Then we used the background gene list for *Homo sapiens* species to conduct the functional annotation. When analyzing the results, we only took the top five annotation cluster groups and extracted annotation information to report all the GO terms for each of the 8 clusters. We then performed GO term comparisons across 8 clusters.

2.3   MiRNA Function and Pathway Analysis

We checked for genes in each cluster to search for pathways associated with breast cancer. We also employed miRbase [10, 11], published articles [12], and DIANA TOOLs [13] to study these miRNAs and their target genes selected from lists enriched with biological processes associated with cancer.

3 Results

3.1 GO Analysis

We found that four breast cancer associated GO terms (Transcription, lumen, nucleolus, and nucleoplasm) are enriched in 4 out of 8 clusters (Fig. 1), followed by nucleoside, nucleotide, and ATP binding that are enriched in 3 data sets. Despite those already listed, there are more GO terms enriched by more than one gene cluster, such as cell cycle, cytoskeleton, chromosomes, non-membrane bounded organelles, microtubules, etc (Fig. 1).

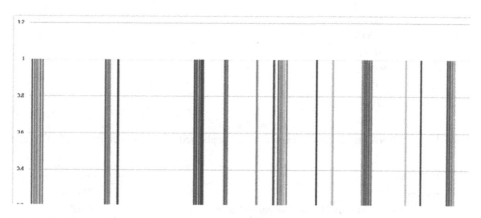

Fig. 1. Summary of the GO terms for 8 cluster genes identified in BRCA (The numbers above cluster name are the number of genes used in DAVID)

Three or more out of the 42 known breast cancer risk genes have been reported in 2 out of all 8 clusters (Table 1). Cluster 3 has the term GO:0007049 $\sim$ cell cycle with a $p < 0.001$, and a well-known breast cancer risk gene BRCA1 found in the cluster. Cluster 6 also has the term GO:0007049 $\sim$ cell cycle with $p < 0.001$. Cluster 7 is reported with *PTEN* and *TP53*, both playing roles in tumor suppression.

Four MiRNAs, *hsa-mir-193b*, *hsa-mir-377*, *hsa-mir-505*, and *hsa-mir-641* are found targeting more than 1 known breast cancer risk genes in C3, C6, and C7. Research showed that *hsa-mir*-193b suppressed the ability of MCF-7 cells' resistance to doxorubicin, an active cytotoxic agent for breast cancer treatment [12].

We have also obtained the miRNA-mRNA pair lists associated with biological processes and summarized the information in Table 2.

Table 1. Summary of known genes implicated with breast cancer found in each cluster

	C1 (572)	C2 (491)	C3 (699)	C4 (513)	C5 (349)	C6 (570)	C7 (489)	C8 (342)
ATM				x				
BRCA1			x					
BRIP1			x					
EPCAM						x		
FANCA						x		
FANCC	x							
FANCD2			x					
FANCF			x					
FANCG						x		
FANCI						x		
MEN1			x					
PALLD				x				
PTEN							x	
TP53							x	

Table 2. MicroRNAs targeting the known genes implicated with breast cancer in C3, C6, and C7.

MiRNAs	Genes
Hsa-mir-758	*BRCA1*
hsa-mir-377, hsa-mir-381, & hsa-mir-193b	*BRIP1*
hsa-mir-193b & hsa-mir-23c	*FANCD2*
hsa-mir-1179, hsa-mir-1262, & hsa-mir-379	*FANCF*
hsa-mir-539, hsa-mir-377, hsa-mir-432, hsa-mir-370, & hsa-mir-495	*MEN1*
hsa-mir-641	*EPCAM*
hsa-mir-452	*FANCA*
hsa-mir-4286 & hsa-mir-185	*FANCG*
hsa-mir-641	*FANGI*
hsa-mir-205, hsa-mir-224, hsa-mir-584, hsa-mir-570, & hsa-mir-505	*PTEN*
hsa-mir-375, hsa-mir-223, & hsa-mir- 505	*TP53*

4 Conclusions

We notice that GO terms are not often shared across clusters. Breast cancer related GO terms (cell cycle) seems to be prevalent in two significant clusters with four or more known breast cancer risk genes. Several BRCA associated genes targeted by MiRNA (s) are identified in our cluster results. BRCA associated genes could be targeted by BRCA MiRNAs (e.g. *hsa-mir-193b*). Many targeting MiRNAs are involved in BRCA pathway (e.g. PI3K-Akt). This implies that both miRNA and gene with their targeting relationship could play important roles in breast cancer oncogenesis.

Acknowledgements. The results presented in this poster here are in whole based upon data generated by the TCGA Research Network: https://www.cancer.gov/types/breast.

References

1. Snyder, M.: Genomics and Personalized Medicine: What Everyone Needs to Know®. Oxford University Press, New York City (2016). Printed by Sheridan, USA
2. Huang, S., et al.: Genes and functions from breast cancer signatures. BMC Cancer **18**(1), 473 (2018). https://doi.org/10.1186/s12885-018-4388-4
3. Ding, L., et al.: Clustering analysis of microRNA and mRNA expression data from TCGA using maximum edge-weighted matching algorithms. BMC Med. Genomics **12**(1), 117 (2019). https://doi.org/10.1186/s12920-019-0562-z
4. Bisognin, A., et al.: MAGIA^2: from miRNA and genes expression data integrative analysis to microRNA-transcription factor mixed regulatory circuits (2012 update). Nucleic Acids Res. **40**(W1), W13–W21 (2012). https://doi.org/10.1093/nar/gks460
5. Da Silveira, W.A., et al.: miRmapper: a tool for interpretation of miRNA-mRNA interaction networks. Genes **9**(9), 458 (2018)
6. Vincent, D.B., Jean-Loup, G., Renaud, L., Etienne, L.: Fast unfolding of communities in large networks. J. Stat. Mech Theory Exp. **2008**, P10008 (2008)
7. Dai, X., Ding, L., Liu, H., Zesheng, X., Jiang, H., Handelman, S., Bai, Y.: Identifying interaction clusters for MiRNA and MRNA pairs in TCGA. Genes **10**(9), 702 (2019)
8. Huang, D.W., Sherman, B.T., Lempicki, R.A.: Systematic and integrative analysis of large gene lists using DAVID bioinformatics resources. Nature Protoc. **4**(1), 44–57 (2009). [PubMed]
9. Huang, D.W., Sherman, B.T., Lempicki, R.A.: Bioinformatics enrichment tools: paths toward the comprehensive functional analysis of large gene lists. Nucleic Acids Res. **37**(1), 1–13 (2009). [PubMed]
10. Kozomara, A., Birgaoanu, M., Griffiths-Jones, S.: miRBase: from microRNA sequences to function. Nucleic Acids Res. **47**, D155–D162 (2019)
11. Kozomara, A., Griffiths-Jones, S.: miRBase: annotating high confidence microRNAs using deep sequencing data. Nucleic Acids Res. **42**, D68–D73 (2014)
12. Long, J., Ji, Z., Jiang, K., Wang, Z., Meng, G.: miR-193b modulates resistance to Doxorubicin in human breast cancer cells by downregulating MCL-1. Biomed. Res. Int. **2015**, 373574 (2015). https://doi.org/10.1155/2015/373574. Epub 2015 Oct 7
13. Vlachos, I.S., et al.: DIANA-miRPath v3. 0: deciphering microRNA function with experimental support. Nucleic Acids Res. 43, W460-W466 (2015). gkv403

Author Index

Printed in the United States
by Bookmasters

Printed in the United States
By Bookmasters